Basics of Wood Anatomy

NIPA GENX ELECTRONIC RESOURCES & SOLUTIONS P. LTD.
New Delhi-110 034

About The Editors

Dr. Kanica Upadhyay did her Ph.D. (Wood science and Technology) in 2016 from Dr. Y. S. Parmar University of Horticulture and Forestry, Nauni, Solan (Himachal Pradesh). She has 6 years of experience in teaching and research field of Forestry. She has done exchange programme with Lakehead University of Natural Sc., Thunderbay, ON (Canada) during M.Sc. and explored new technologies in Canada. She received DST-INSPIRE fellowship for her Ph.D. programme. She started her career as Teaching Associate during Ph.D., then started working as Assistant Professor (Forestry) at Himgiri Zee University, Dehradun in 2017. Currently, she is working as Assistant Professor (Forest Products & Utilization) in College of Horticulture & Forestry, Jhalawar, Agriculture University, Kota, Rajasthan from 2018 onwards. She is actively engaged in teaching & research of UG & PG. Also, she is guiding PG students, actively engaged various extension activities of the College & KVK. She is also a life member of various agriculture societies. She authored 25 peer-reviewed research/review papers, and 03 practical manuals.

Dr. Rajneesh Kumar (Ph.D., Forestry), currently working as Assistant Professor in the Department of Forest Products, Dr YS Parmar University of Horticulture and Forestry, Nauni, Solan (HP).His main area of work includes teaching at Undergraduate and Postgraduate levels, besides research and extension in the field of forest products. He has his expertise in various fields viz., Resin Tapping Techniques, Wood Science & Technology and Agro-techniques of Medicinal and Aromatic plants, etc. He has standardized the Borehole Method of oleoresin tapping in Himalayan Blue Pine (*Pinus wallichiana*). He has guided 05 M.Sc. students besides handling Adhoc projects. He has published about 30 research papers and articles in different national and international journals,01 book chapter,01 Teaching/Practical manual,04 brochures/ bulletins, etc. Presently he is engaged in the field of forestry and his main area of work includes teaching, research and extension

Dr. Sneha Dobhal, has 7 years of teaching and research experience in the field of forestry. She graduated in biotechnology from Sardar Bhagwan Singh P.G.I. of Biomedical Science & Research, Dehradun in 2009, and did her post graduation from Forest Research Institute, Dehradun in 2011 and doctorate in Forestry from Dr. Y.S. Parmar University of Horticulture & Forestry, Nauni, Solan (HP). Currently, she is working as Assistant Professor (Forestry) in VCSG Uttarakhand University of Horticulture and Forestry, College of Forestry, Ranichauri (Uttarakhand), India. She has also worked at ICAR: Indian Institute of Soil & Water Conservation, Dehradun as Junior Research Fellow (JRF). She has edited one book, and published 19 research publications in referred national and international journals. She has coordinated training on "Remote Sensing and GIS Applications in Carbon Forestry" Conducted by Indian Institute of Remote Sensing (IIRS). She has also attended many trainings and workshops. She has won 2 national awards in conferences. She is reviewer member of current science journal (NAAS rating 6.84) and life member of International Journal of Farm Sciences having journal NAAS rating is 4.01.

Basics of Wood Anatomy

Kanica Upadhyay
Assistant Professor
Department of Forest Products and Utilization
College of Horticulture and Forestry, Jhalawar
Agriculture University, Kota
Rajasthan

Rajneesh Kumar
Assistant Professor
Department of Forest Products
Dr. Y.S. Parmar University of Horticulture & Forestry
Nauni, Himachal Pradesh

Sneha Dobhal
Assistant Professor
Department of Silviculture
College of Forestry
Veer Chandra Singh Garhwali Uttrakhand
University of Horticulture & Forestry
Uttrakhand

NIPA GENX ELECTRONIC RESOURCES & SOLUTIONS P. LTD.
New Delhi-110 034

NIPA GENX ELECTRONIC
RESOURCES & SOLUTIONS P. LTD.

101,103, Vikas Surya Plaza, CU Block
L.S.C.Market, Pitam Pura, New Delhi-110 034
Ph : +91 11 27341616, 27341717, 27341718
E-mail:newindiapublishingagency@gmail.com
www: www.nipabooks.com

For customer assistance, please contact
Phone: + 91-11-27 34 17 17
Fax: + 91-11- 27 34 16 16
E-Mail: feedbacks@nipabooks.com

ISBN: 978-93-91383-53-4

Composed and Designed by NIPA.

Preface

Wood anatomy, the study of woody cells and tissues, has advanced considerably since the early descriptive accounts were made which consisted mainly of cataloguing what was 'out there'. Anatomical data have been applied in better understanding of the interrelationships of woody plants, confirming evidence of natural relationships of plant families in combined analyses. This book will serve its purpose well, for undegraduates of forestry. We realized that with the passage of time, many new disciplines had been developed, and older ones expanded to a point where a much revised and updated book of this type could play an important part. Wood is composed mostly of hollow, elongated, spindle-shaped cells that are arranged parallel to each other along the trunk of a tree. The characteristics of these fibrous cells and their arrangement affect strength properties, appearance, resistance to penetration by water and chemicals, resistance to decay, and many other properties. Just under the bark of a tree is a thin layer of cells, not visible to the naked eye, called the cambium. Here cells divide and eventually differentiate to form bark tissue to the outside of the cambium and wood or xylem tissue to the inside. This newly formed wood (termed sapwood) contains many living cells and conducts sap upward in the tree. Eventually, the inner sapwood cells become inactive and are transformed into heartwood. This transformation is often accompanied by the formation of extractives that darken the wood, make it less porous, and sometimes provide more resistance to decay. The center of the trunk is the pith, the soft tissue about which the first wood growth takes place in the newly formed twigs.

The authors hope that this book will be helpful in imparting theoretical skills to the students, academicians and teaching faculty of the forestry and agricultural disciplines working in field of woody plants. This book is intended to impart basic education for the UG students of forestry for the course wood anatomy.

Kanica Upadhyay
Rajneesh Kumar
Sneha Dobhal

Contents

Preface .... *vii*

1. Introduction to Plant Anatomy 1
2. Classification of Plant Kingdom 5
3. Plant Body Development 13
4. Plant Tissues 17
5. Anatomy of Stem 31
6. Anatomy of Root 39
7. Secondary Growth 45
8. Tissue System 53
9. Macroscopic Features of Wood 65
10. Microscopic Features of Wood 71
11. Physical Properties of Wood 77
12. Chemical Composition of Wood Cell Walls 85
13. Anatomical Features of Common Indian Timbers 87

Appendix *135*

1

Introduction to Plant Anatomy

Anatomy

The study of gross internal structure of plant organs by the technique of section cutting is called anatomy .The term anatomy was derived from two words **'ana'** means as under and **'temnein'** meaning to cut. Anatomy is the branch of science which deals with internal structures and organization of the organism. It is also known as internal morphology. Plant anatomy also describes tissue and tissue systems of the plant body.

History/Introduction of Plant Anatomy

Date	Investigator	Major Contributions	Concepts
369 -262BC	**Theophrastus of Eresus** "Father of Botanical Science" Greek Philosopher	Descriptive morphology of types of organs, relation of organ to organ, and of type of organ to type of organ. Described gross internal anatomy of stems, roots, and leaves.	Recognized normal sequence of root, stem, branch, leaf, flower, and fruit in trees. Plants are made up of bark (phloios), wood (zylon), and pith (metra), when pith is present.
1519 - 1603	Andrea Caesalpino Italian Philosopher	Initiated idealistic morphology in studies on plant "souls".	Postulated canals for conduction. Observed that roots lack in pith.
1635 - 1703	Robert Hooke english mathematician & architect	Examined many objects with newly invented magnifying lens. (*Micrographica.*, 1665).	Cell coined in reference to cavities in cork cambium and charcoal.
1628 - 1694	Marcello Malpighi Italian physician and professor	Searched independently of Grew for similarities between structures of animals and plants.	Discovered Spiral Vessels and Stomata.

Date	Investigator	Major Contributions	Concepts
1641 - 1712	Nehemia Grew "Founder of Plant Anatomy" English Physician	Presented classification of plant tissues as consisting of two different "bodies" (woody parts, strings, and fibers vs. barques, piths, parenchymas, and pulps). Recognized vertical and horizontal systems of tissues. Described secondary growth of bark and wood. The Anatomy of Vegetables Begun, 1672.	"...Parenchyma of the Barque is much the same thing, as to its conformation, which the Froth of Beer or Eggs is, as a fluid..." Vessel coined in reference to spiral vessels.
1632 - 1723	Antony Van Leeuwenhoek Dutch	-	Described pitted vessels.
1700 - 1781	Du Hamel French Arboriculturist	-	Coined Cambium in reference to gelatinous generative zone in the inner cortex.
1733 -1794	Caspar Friedrich Wolff German	Theorized that tissue is a homogeneous matrix filled with bubbles, as is rising dough.	
1776 - 1854	Charles Francois Mirbel French	Elaborated Wolff Theory. Theorized that new cells appear in a homogeneous matrix as cavities with openings between them for the passage of sap.	
1766 - 1833	Kurt Sprengel	Opposed Mirbel's Theory. Proposed new cells arise within the contents of old cells as small vesicles (starch grains) that get larger by uptake of water.	
1779 - 1864	Ludolph Christian Treviranus	Discovered vessels formed by disappearance of cross-walls between series of cells. Observed development of spiral thickenings in protoxylem.	
1774 - 1850	Johann Jakob Bernhardi	Discovered annular thickenings in vessels. Observed primary wall binding annular and spiral thickenings together. Recognized that vessel elements don't metamorphize.	
1766 - 1827	Johann Jakob Paul Moldenhawer	Demonstrated each cell had its own wall, so cavities are separated by two walls.	
1773 - 1858	Robert Brown	Cell Anatomy	Discovered cell nucleus (1831)
1804 - 1840	Franz Julius Ferdinand Meyen	Suggested new cell arise through cell division rather than free cell formation.	

Date	Investigator	Major Contributions	Concepts
1805 -1872	Hugo von Mohl	Proposed that the vacuolated body (primordial utricle) within the cell wall was the living component. Proposed cell wall thickening occurs by apposition.	Coined term protoplasm in reference to cell contents. Described relation of primary and secondary cell wall layers and nature of pits.
1805 - 1880	Theodor Hartig		Discovered sieve tube and its perforated nature in phloem.
1804 -1881	Matthias Jacob Schleiden	Studied protoplasm of plants.	
1810 - 1882	Theodor Schwann	Studied protoplasm of animals.	The "Cell theory (1838)" there is one universal principle of development for the elementary parts of organisms, however different, and that this principle is the formation of cells.
1817 - 1891	Wilhelm Von Nageli	Studied ontogeny of apical meristems. Distinquished primary and secondary meristems. Proposed cell wall thickening occurs by intussusception.	Described development of vascular bundles from procambial strands. Applied terms xylem and phloem to different parts of vascular bundle.
1832 - 1891	Carl Sanio Prussian, school teacher	Described how vascular cambium originates and functions. Described secondary development of periderm.	Described details of bordered pits.
1822 - 1880	Johannes Von Hanstein	Proposed theory of histogen apical meristem organization.	Dermatogen Periblem. Cal
1831 - 1888	Heinrich Anton De Bary Mycologist	Demonstrated Hanstein's histogens could not be universally applied and lacked morphological value.	
1832 - 1897	Julius Von Sachs	Proposed first physiological classification of plant tissue that were derived from uniform meristem.	Epidermal. Fibrovascular. Fundametal.

Date	Investigator	Major Contributions	Concepts
1884	Gottlieb Haberlandt	Physiologische Pflanzenanatomie. grouped tissues according to functional systems, disregarding morphological classification and arrangements.	
1924	Schmidt	Proposed method of apical organization on basis of planes of cell division.	Tunica. Corpus. Kappe.
1943	Foster	Proposed zones of growth description of gymnosperms apical meristems.	Apical Initials. Mantle Layer. Central Mother Cell Zone.
1961	Clowes	Described quiescent center in root apical meristems.	

2

Classification of Plant Kingdom

Arrangement of plants and animals into groups according to their close resemblance, naming and describing them is called classification. Classifying the plants into groups is called systematic botany. The main aim of classification is to name the plant to describe them and to bring out their resemblances in order to throw light on their relationship and common ancestor. It is to indicate primitive and advanced plants systematically.

Need for Classification

It is not possible for anyone to study all the organisms. But if they are grouped in some convenient way the study would become easier as the characters of a particular group or a family would apply to all the individuals of that group. Classification allows us to understand diversity better.

Classification of plants

Entire plant kingdom is divided into four main groups.

1. Thallophyta 2. Bryophyta 3. Pteridophyta 4. Spermatophyta or Phenerogams

1. Thallophyta

These are simplest and primitive unicellular or filamentous (thread like) plants. They are green as well as non green and their body is not differentiated into root stem and leaves. Such a plant body is called thallus and hence is the name of the group. It consists of three sub-groups: -

a) Bacteria b) Algae c) Fungi

A) Bacteria

These are minute microscopic unicellular, non-green round or rod like structures. They are either saprophytes or parasites. They are harmful because they cause many diseases of plants and animals. They are useful also in industries such as alcohol and vinegar production in tea, leather and dairy industries. Reproduction is by spores e.g. *Bacillus* (Hay Bacteria) and *Lactic* Bacteria.

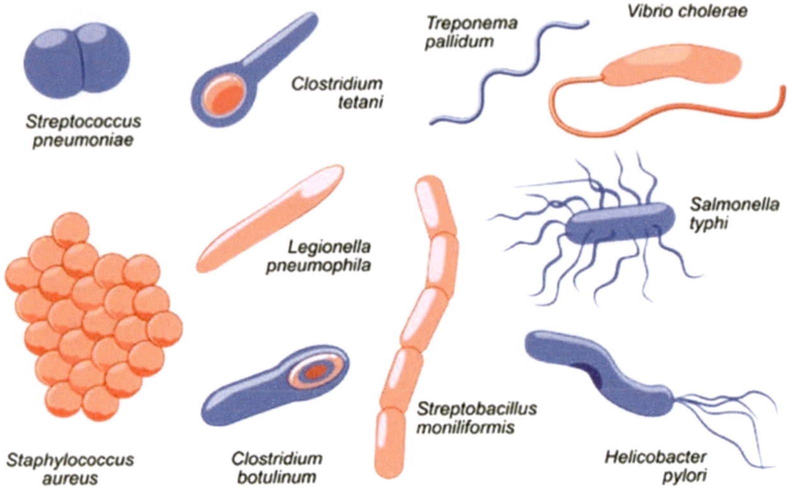

Fig 2.1. Bacteria

B) Algae

Unicallular (e. g. *Chlamydomonas*) or filamentous (e.g. *Spirogya*) and ribbon like, many hundred feet in length (e.g. *Laminaria*). Algae may be green, brown or red. Reproduction is vegetative as well as sexual.

Fig. 2.1.2. Algae

C) Fungi

These are all non green plants and their thallus is called *mycellium* consisting of white threads called *hyphae.* Each hypha is a tube with protoplasm, vacuoles and many nuclei and oil drops. There is no chlorophyll. Mode of nutrition is *saprophytic* and thallus is saprophyte, i.e. it absorbs food material in solution form by its hyphae from dead and decomposed organic remains of animals and vegetables. Reproduction is a asexual by spores and sexual by *gametes. Rhizopus, mucor*. Some higher forms of fungi grow erecton soil with big stemlike structure. They are edible also e.g. *Morchella* (Guchhi) and *Agaricus* (Khumb).

Fig. 2.1.3. Fungi

2. Bryophyta

They are land plants and plant body is thallus and green in colour being rich in chloroplasts. They are moisture and shade lovers and are found attached to substratum (rocks and banks of streams) by thread like structures called rhizoides. It has following two sub-groups.

a) Liverworts

They have flat thallus like the palm of hand. Dorsal side is green with vegetatively ventral side is pale yellow with rhizoids attached to substratum e.g. *Marchantia* and *Riccia*.

Fig. 2.2.4. Bryophyta

b) Mosses

They are plants with errect green leafy shoots. They can tolerate drought very well. Separate male and female shoots are present on the same plant. Reproduction is vegetative as well as sexual e.g. *Funaria*

Fig.2.2.5. Mosses

3. Pteridophyta

These are land plants with well developed roots, stem and leaves. They are also shade and moisture lovers. They have tracheids for conduction of water. Some have underground stem called rhizome, others have erect aerial stem as in ferns. Many have stem creeping above the soil e.g. *Selaginella.* Some are aquatic plants e.g. water fern (*Marsilea*). Reproduction is vegetative by tubers as well as sexual. All the above mentioned three groups of plants are also called Cryptogams or Non Flowering Plants.

Fig. 2.3. Pteridophytes

4. Spermatophyta or Phenerogams

This group of plants is very advanced group than all the non-flowering plants or crytograms because they bear flower and reproduce by seeds. They may be naked or enclosed in ovary. Vascular systems are advanced and contain well developed xylem and phloem. Xylem has tracheids (Gymnosperm) and vessels (Angiosperm). They all have definite root stems and leaves and occur in different habitats. They are herbs, shrubs, or tall trees. Methods of pollinations and sexual reproduction are complicated and advanced. They are divided into two groups.

a) Gymnosperm

The gymnosperms (*gymnos : naked, sperma: seeds) are plants in which* the ovules are not enclosed by any ovary wall and remain exposed, both before and after fertilization. The seeds that develop post-fertilization, are not covered, i.e., are naked. Gymnosperms include medium-sized trees or tall trees and shrubs. One of the gymnosperms, the giant redwood tree *Sequoia is one of the tallest tree species. The roots are* generally tap roots. Roots in some genera have fungal association in the form of **mycorrhiza *(Pinus), while in some others (Cycas) small specialised*** roots called coralloid roots are associated with N_2 fixing cyanobacteria. The stems are unbranched (*Cycas) or branched (Pinus, Cedrus). The leaves* may be simple or compound. In *Cycas the pinnate leaves persist for a few* years. The leaves in gymnosperms are well-adapted to withstand extremes of temperature, humidity and wind. In conifers, the needle-like leaves reduce the surface area. Their thick cuticle and sunken stomata also help to reduce water loss. The gymnosperms are heterosporous; they produce haploid microspores and megaspores. The two kinds of spores are produced within sporangia that are borne on sporophylls which are arranged spirally along an axis to form lax or compact strobili or cones. The strobili bearing microsporophylls and microsporangia are called microsporangiate or male strobili.

Fig. 2.4.1. Gymnosperms

The microspores develop into a male gametophytic generation which is highly reduced and is confined to only a limited number of cells.This reduced gametophyte is called a **pollen grain. The development of pollen grains takes place within** the microsporangia. The cones bearing megasporophylls with ovules or **megasporangia are called** macrosporangiate or **female strobili. The male or female** cones or strobili may be borne on the same tree (*Pinus).* However, in *cycas* male cones *and megasporophylls are* borne on different trees. The megaspore mother cell is differentiated from one of the cells of the nucellus. The nucellus is protected by envelopes and the composite structure is called an **ovule. The ovules are borne on** megasporophylls which may be clustered to form the female cones. The megaspore mother cell divides meiotically to form four megaspores. One of the megaspores enclosed within the **megasporangium** develops into a multicellular female gametophyte that bears two or more **archegonia or female sex organs. The** multicellular female gametophyte is also retained within megasporangium. Unlike bryophytes and pteridophytes, in gymnosperms the male and the female gametophytes do not have an independent free-living existence. They remain within the sporangia retained on the sporophytes. The pollen grain is released from the microsporangium. They are carried in air currents and come in contact with the opening of the ovules borne on megasporophylls. The pollen tube carrying the male gametes grows towards archegonia in the ovules and discharges their contents near the mouth of the archegonia. Following fertilization, zygote develops into an embryo and the ovules into seeds. These seeds are not covered.

b) Angiosperms

Unlike the gymnosperms where the ovules are naked, in the angiosperms or flowering plants, the pollen grains and ovules are developed in specialised structures called flowers. In angiosperms, the seeds are enclosed by fruits. The angiosperms are an exceptionally large group of plants occurring in wide range of habitats. They range in size from tiny, almost microscopic *Wolfia* to tall trees of *Eucalyptus* (over 100 metres). They provide us *with* food, fodder, fuel, medicines and several other commercially important products. They are divided into two classes: the dicotyledons and the monocotyledons. The dicotyledons are characterised by having two cotyledons in their seeds while the monocolyledons have only one. The male sex organ in a flower is the stamen. Each stamen consists of a slender filament with an anther at the tip. The anthers, following meiosis, produce pollen grains.

Fig.2.4.2. Angiosperms

The female sex organ in a flower is the pistil or the carpel. Pistil consists of an ovary enclosing one to many ovules. Within ovules are present highly reduced female gametophytes termed **embryosacs.** The embryo-sac formation is preceded by meiosis. Hence, each of the cells of an embryo-sac is haploid. Each embryo-sac has a three-celled egg apparatus – one egg cell and two synergids, three antipodal cells and two polar nuclei. The polar nuclei eventually fuse to produce a diploid secondary nucleus. Pollen grain, after dispersal from the anthers, is carried by wind or various other agencies to the stigma of a pistil. This is termed as pollination.

The pollen grains germinate on the stigma and the resulting pollen tubes grow through the tissues of stigma and style and reach the ovule. The pollen tubes enter the embryo-sac where two male gametes are discharged. One of the male gametes fuses with the egg cell to form a zygote (syngamy). The other male gamete fuses with the diploid secondary nucleus to produce the triploid primary endosperm nucleus (PEN). Because of the involvement of two fusions, this event is termed as double fertilization, an event unique to angiosperms. The zygote develops into an embryo (with one or two cotyledons) and the PEN develops into endosperm which provides nourishment to the developing embryo. The synergids and antipodals degenerate after fertilization. During these events the ovules develop into seeds and the ovaries develop into fruit.

3

Plant Body Development

Development of Plant Body

A vascular plant begins in existence as a morphologically simple unicellular zygote (2n). The zygote develops into the embryo which in the long run develops into the mature sporophyte. The development of the sporophyte (2n) involves division and differentiation of cells and organization of cells into the tissue systems. The embryo of the seed plant possesses relatively simple structure as compared to the mature sporophyte. The embryo bears a limited number of parts containing generally only axis bearing one or more cotyledons. The cells and the tissues of this structure are less differentiated. However, the embryo grows further, because of the presence of the meristems, at two opposite ends of the axis, of future shoot and root. After germination of the seed during development of the shoot and root new apical meristems appear which cause a vegetative growth and reproductive stage of the plant attained.

Fundamental Parts of the Plant Body

The plant body consists of a number of organ *i.e.,* root, stem, leaf and flower. Each part has different functions for plant growth.Flower consists of sepal, petals, stamens, carpels and sometimes also sterile members. Each organ is made up of number of tissue systems; each tissue system consists of number of tissues. Tissue consists of group of cells that are alike in origin, structure and function.

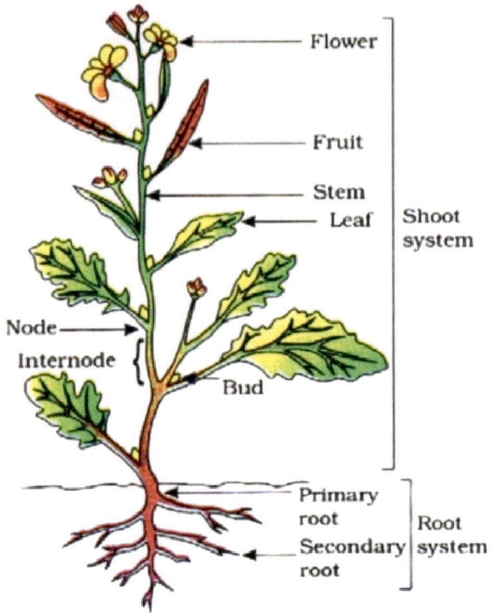

Fig. 3.1. Fundamental Parts of the Plant Body

The plant body consists of two main systems namely, root system and shoot system. Both of these systems have a common axis. Root system is generally underground and positively geotropic. Roots are non-green brown in colour. They are not differentiated into nodes and internodes. The chief functions of root system includes fixation of the plant in soil, absorption of mineral water and conduction of mineral water to the root system. Shoot system is aerial part of the plant and it is negatively geotropic. Main axis of the shoot system is called stem. The stem bears branches, leaves, buds and flower, etc. The stem is distinguished into nodes and internodes. The leaves develop over the stem at the nodes only. The part of the stem between two successive nodes is called internode. The main stem and its branches end with terminal bud or apical bud. The main functions of shoot system includes display the foliage by branches, conducts mineral water from the root system and food material to the root system, and gives mechanical strength to the plant body.

Growth of the Plant Body

The growth of plant occurs from the embryo is called primary growth. The tissues formed from the embryo are called primary tissues. Example: primary xylem. Generally constant growth in monocots occurs due to the primary growth. Secondary growth is absent in monocots. In the most vascular cryptogams and monocotyledons, the entire life cycle of the sporophyte is completed in primary plant body. The gymnosperms, most dicotyledons and some monocotyledons show an increase in thickness of the stem and root by means of secondary growth. The tissues formed as a result of secondary growth are called secondary tissues. A special meristem, the cambium, is concerned with the secondary thickening. The cambium arises between the primary xylem and the primary phloem, and lays down the new xylem and phloem adjacent to these. The bulk of the plant increases because of secondary growth. Especially the vascular tissues are developed which provide new conducting calls and additional support and protection. In addition, a cork cambium or phellogen commonly develops in the peripheral region of the axis and produces a periderm in the stem, a secondary tissue system assuming a protective function when the primary epidermal layer is disrupted during the secondary growth in thickness. The primary growth of an axis is completed in relatively short period, whereas the secondary growth persists for a longer period, and in a perennial axis the secondary growth continues indefinitely. The plant body consists of a number of small, microscopic, boxes like compartments called cells. Fundamental basis of the life is cell.

Kinds of Woody Plants

A woody plant is a plant that produces wood as its structural tissue. Woody plants are usually either **trees**, **shrubs**, or **lianas**. These are usually **perennial** plants whose stems and larger roots are reinforced with wood produced from secondary xylem.

1) Trees

A woody perennial plant, typically having a single stem or trunk growing to a considerable height and bearing lateral branches at some distance from the ground.

Fig.3.2. Tree

2) Shrubs

Woody plants without a single main stem, occurring well below the forest canopy. For Examples: *Lantana camara, Daphanae cannabina, Carissa carandus, Woodfordia* spp., etc.

Fig.3.3. Shrub

3) Lianas

A **liana** is any of various long-stemmed, woody vines that are rooted in the soil at ground level and use trees, as well as other means of vertical support, to climb up to the canopy to get access to well-lit areas of the forest. Lianas are characteristic of tropical moist deciduous forests (especially seasonal forests), but may be found in temperate rainforests. There are also temperate lianas, for example *Clematis or vitis, Calamus* spp. *Capparis* spp., *Dioscorea* spp.

Fig.3.4. Lianas

4

Plant Tissues

Meristematic Tissues

These are composed of cells that are in state of division or retain the power of dividing. These cells are essentially alike and isodiametric. They may be spherical, oval or polygonal and their walls thin and homogeneous; the protoplasm in them abundant and active with large nuclei, and the vacuoles small or absent.

Classification of Meristems

Meristems are classified in different ways on the basis of certain factors.

A) **According to their origin and development**

1. Promeristem or primordial meristem
2. Primary meristem and
3. Secondary meristem.

B) **According to their position in the plant body**

1. Apical
2. Intercalary and
3. Lateral meristem.

C) According to their functions

1. Protoderm
2. Procambium
3. Ground / fundamental meristem.

A) According to their origin and development

1. **Promeristem or primordial meristem:** These consists of a group of meristematic cells representing the earliest or youngest stage of a growing organ. It is, in fact the stage from which differentiation of later

meristems and finally, of permanent tissue, takes place. It occupies a small area at the tip of the stem and the root. The **promeristem or primordial meristem** gives rise to the primary meristem by cell division and is therefore the earliest stage or originator of the latter.

2. **Primary meristem:** It is derived from the **promeristem or primordial meristem,** and still fully retains its meristematic activity. As a matter of fact, its cells divide rapidly and become differentiated into distinct tissues – the primary permanent tissues which make up the fundamental structure of the plant body. It is primarily the growing apical region of the root and stem. It should also be noted that the cambium of the stem is also a primary meristem although it gives rise to the secondary permanent tissues. Another fact to be noted in this connection is that the cambium cells divided in one plane (tangential), while those of the primary meristem divide in 3 planes.

3. **Secondary meristem:** These appear later, at a certain stage of the development of an organ of a plant. It is always lateral, lying along the side of the stem and root. The primary permanent tissues become meristematic, i.e. they acquire the power of division and constitute the secondary meristem e.g. the cambium of root, the interfascicular cambium of the stem and the cork – cambium of both. All lateral meristems (primary & secondary) give rise to the secondary permanent tissues and are responsible for growth in the thickness of the plant body.

B. According to their position in the plant body

1. **Apical meristem :** Apical meristem is found at the tips of roots, stem and branches.It includes the promeristem and the primary meristem, to give rise the primary permanent tissues and is responsible for increase in length of plant body. It is divided into three zones – protoderm, procambium and ground meristem. Protoderm gives rise to epidermal tissue; procambium gives rise to primary vascular tissues and ground meristem gives rise to cortex and pith

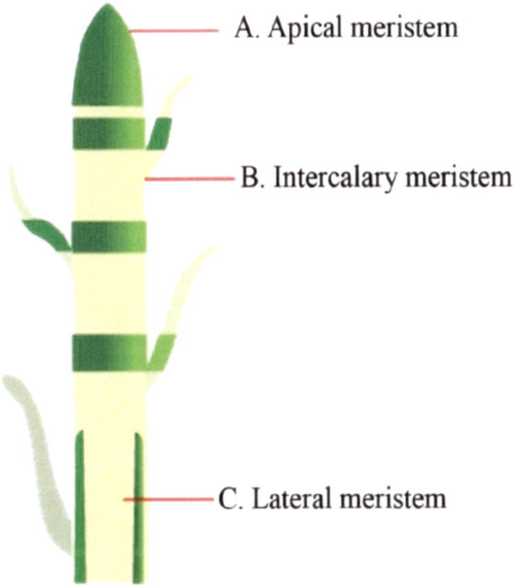

Fig.4.1. Schematic representation of position of different meristems

2. **Intercalary meristem:** It is present in the nodal region and is prominently found in monocotyledons, *e.g.* grasses. Sometimes below the node, as in mint (*Mentha*). It is detached portion of the apical meristem, separted from the latter due to growth of the organ. As the name indicates, it is present in between the permanent tissues. It is derived from the apical meristem and is responsible for the elongation of internodes. It is generally short – lived, either disappearing soon or becoming transformed into permanent tissues.

3. **Lateral meristem:** The meristem that is present along the longitudinal axis of stem and root is called lateral meristem. Vascular cambium and cork cambium (phellogen) are examples for lateral meristem. It divides mainly in the tangential direction and giving rise to the secondary permanent tissues to the inside and outside of it, which result in the thickening of stem and root.

C. Haberlandt in 1914, gave classification of meristem according

1. **Protoderm :** Give rise to the epidermal tissue system
2. **Procambium :** Give rise to vascular tissue system
3. **Ground or fundamental meristem:** Ground tissue system

Permanent tissues

The cells, which are formed by apical meristem, are differentiated into different types of permanent tissues. These tissues have lost the power of dividing either permanently or temporarily.

Classification of permanent tissue

Based on the constituent cells, the permanent tissue is classified into two types:

1. Simple tissue 2. Complex tissue.

1. Simple Tissue

A tissue with the cells of similar structure and function is called simple tissue. It is of three types **:**

A. Parenchyma

B. Collenchyma

C. Sclerenchyma

Parenchyma

It is generally present in all organs of the plant. It constitutes the ground tissue in a plant. Parenchyma is the precursor of all the other tissues. Its main function is to storage of food. Parenchyma is a living tissue and made up of thin walled cells. The cell wall is made up of cellulose. Parenchyma cells may be oval, spherical, rectangular, cylindrical or stellate. The cells are usually polyhedral with 10 to12 facets. Parenchyma is of different types and some of them are discussed as follows.

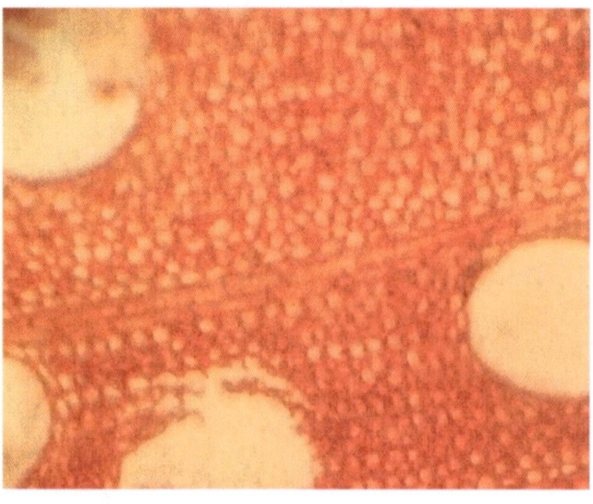

Fig.4.2. Parenchyma Cells

In water plants, the parenchyma found in the cortex region possesses well-developed large intercellular spaces called air spaces. This air filled parenchyma tissue is called **Aerenchyma.** It helps the plant to float in water. eg. *Nymphaea* and *Hydrilla*. The parenchyma cells that are stored with starch grains are called **Storage Parenchyma**. eg. stem and root tubers. In the petioles of Banana and Canna, star shaped parenchyma cells are found. These cells are called **Stellate Parenchyma**. In green parts of the plants, the parenchymatous cells have chloroplasts. These cells are called **Chlorenchyma**. Its important function is photosynthesis.

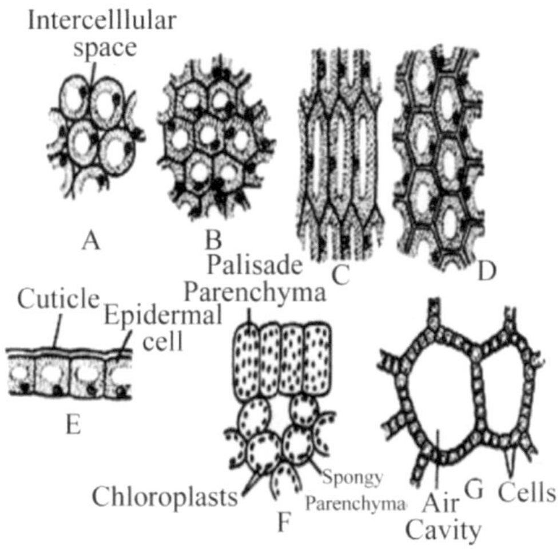

Fig. 4.3: Types of Parenchyma Cells: A-B, Normal parenchyma cells;A. Rounded; B. Angular; C. Prosenchyma; D. Xylem parenchyma; E. Epidermal cells; F. Mesophyll; G. Aerenchyma

Collenchyma

Collenchyma generally occurs in the dicot stems in two or more layers below the epidermis. These layers constitute the hypodermis. It is absent in the roots of land plants. It also occurs in petiole and pedicel.It gives strength to young organs. Collenchyma is a living tissue. It consists of more or less elongated cells, which are polygonal in cross section. The cell wall is unevenly thickened. The thickening is confined to the corners of the cells. Besides cellulose, the cell wall contains high amounts of hemicellulose and pectin. Collenchyma may contain chloroplasts and carry out photosynthesis.

Collenchyma is divided into three types – lamellar, angular and lacunate collenchyma. Its functions are, therefore, both mechanical and vital.In the hypodermis of Helianthus, only the tangential walls of collenchyma are thickened and the radial walls are devoid of thickening. This type of collenchyma is called lamellar collenchyma. In the hypodermis of Datura and Nicotiana, the cell walls of collenchyma are thickened at their angles. This type is called angular collenchyma.In the hypodermis of Ipomoea, the cell wall thickening materials are deposited on the walls bordering the intercellular spaces. This type is called lacunate collenchyma.

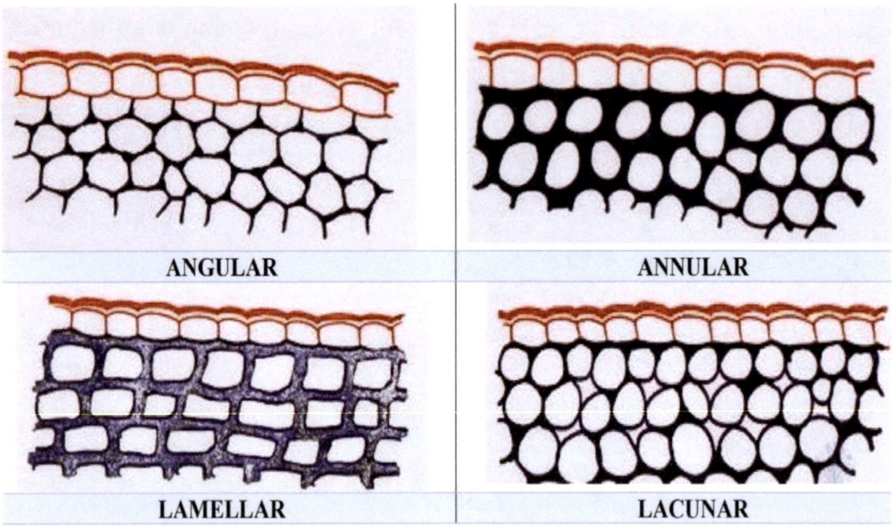

Fig. 4.4: Types of Collenchyma

Sclerenchyma

Sclerenchyma is a dead tissue. The cells have lignified secondary walls. They lack protoplasts. On the basis of origin, structure and function, sclerenchyma is divided into two types – **sclereids and fibres.** The sclereids are different from fibres in the following respects. Sclereids are shorter whereas fibres are longer. Sclereids possess numerous pits as compared to the fibres.

Sclereids

Sclereids are dead cells. They vary greatly in shape and thickness. The cell wall is very thick due to lignification. Lumen is very much reduced. The pits may be simple or branched.Usually sclereids are isodiametric, but in some plants they are elongated. They are responsible for the rigidity of the seed-coat.

Types of Sclereids

a. Brachysclereids

The isodiametric sclereids are called **brachy-sclereids** (stone cells). They are found in bark, pith, cortex, hard endocarp and fleshy portions of some fruits. eg. pulp of *Pyrus*.

Fig.4.4.1. Brachyselereids

b. Macrosclereids

Elongated rod shaped sclereids are called **Macrosclereids** (rod cells). They are found in the outer seed coat. eg. Crotalaria.

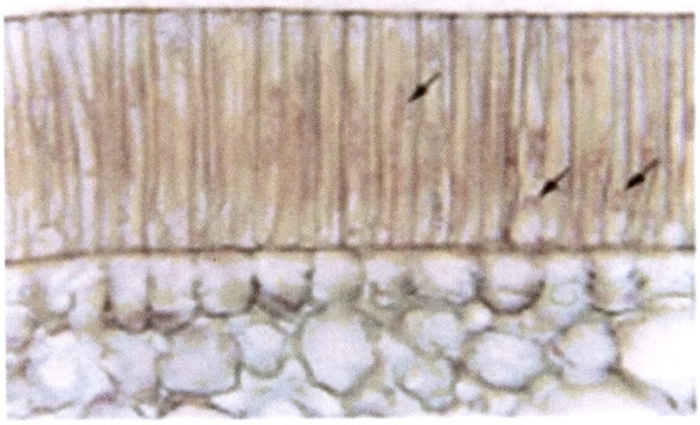

Fig. 4.4.2. Macrosclereids

c. Osteosclereids

The rod shaped sclereids with dilated ends are called Osteosclereids (bone cells). eg. Seed coat of *Pisum* (Pea).

Fig. 4.4.3. Osteosclereids

d. Astrosclereids

They are irregular stone cells, i.e. they are branched in an irregular way, with radiating arms of varying lengths, giving a stellate or star –like appearance. They are found in leaves of certain dicotyledons, as in tea leaf and also in *Tsuga* and *Gnetum.* They also found in the bark of certain conifers, as in *Abies* and *Larix*.

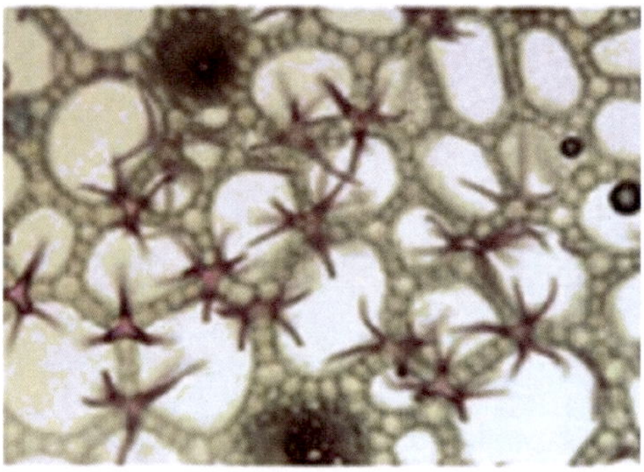

Fig. 4.4.4. Astrosclereids

Fibres

Fibre cells are dead cells. They are very long and narrow with pointed ends. In transverse section, the fibres are polygonal with narrow lumen. The secondary wall is evenly thickened with lignin. It possesses simple pits. Fibres are supporting tissues. They provide mechanical strength to the plants and protect them from the strong winds. The fibres that are found in the seed coat of some seeds are called surface fibres. eg. Cotton.

Comparison: Fibres and Sclereids

Fibre	Sclereids
Elongated and narrow like thread	Usually broad
End walls tapering	End walls blunt is unbranched
Usually unbranched	May be branched
Pits narrow and unbranched	Pits deep and commonly branched
Fibres are formed directly from derivatives of meristematic cells	Sclereids arise by secondary thickening of parenchyma cells

Complex Tissues

A tissue that consists of several kinds of cells but all of them function together as a single unit is called complex tissue. It is of two types :

A. Xylem

B. Phloem

A. Xylem:

Xylem (Greek word 'xylos'= wood) is a complex tissue that is mainly responsible for the conduction of water and mineral salts from roots to other parts of the plant. The xylem, which is derived from procambium, is called primary xylem and the xylem, which is derived from vascular cambium, is called secondary xylem.Earlier formed xylem elements are called protoxylem, whereas the later formed xylem elements are called metaxylem.

Xylem is made up of four kinds of cells:

a. Tracheids

b. Vessels or tracheae

c. Xylem fibres

d. Xylem parenchyma

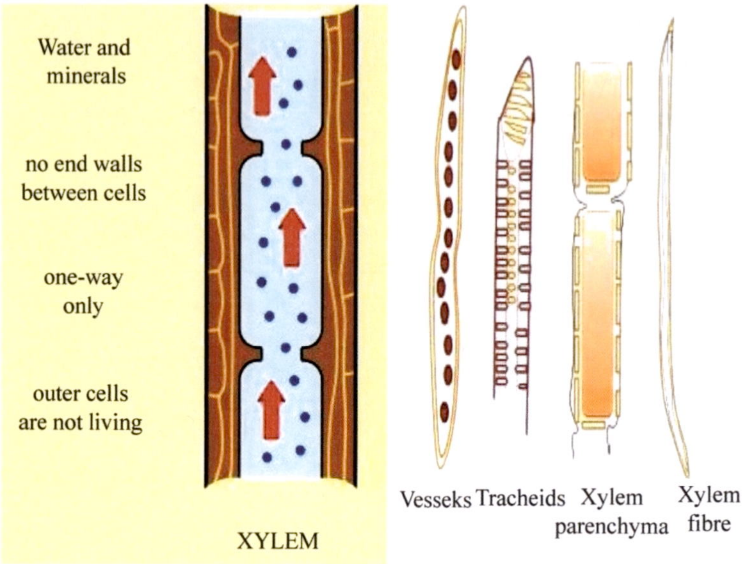

Fig. 4.5.1. Xylem and its elements

a. Tracheids

Tracheids are elongated with blunt ends. Its lumen is broader than that of fibres. Their secondary wall is lignified. In cross section; the tracheids appear polygonal and thick walled. The pits are simple or bordered. There are different types of cell wall thickening due to deposition of secondary wall substances. They are annular (ring like), spiral (spring like), scalariform (ladder like), reticulate (net like) and pitted (uniformly thick except at pits). Tracheids are imperforate cells with bordered pits on their end walls. They are arranged one above the other. Tracheids are chief water conducting elements in gymnosperms and pteridophytes. Here, the conduction of water and mineral salts takes place through the bordered pits. They also offer mechanical support to the plants.

b. Vessels or Tracheae

Vessels are perforated at the end walls. Its lumen is wider than that of tracheids. The perforated plates at the end wall separate the vessels. They occur parallel to the long axis of the plant body. Due to dissolution of entire end wall, a single pore is formed at the perforation plate. It is called simple perforation plate eg. *Mangifera*. If the perforation plate has many pores, then it is called multiple perforation plate eg. *Liriodendron*. The secondary wall thickenings of vessels are annular, spiral, sca-lariform, reticulate, or pitted as in tracheids. Vessels are chief water conducting elements in angiosperms and they are absent in pteridophytes and gymnosperms. However, in Gnetum of gymnosperms, vessels occur.The main function of vessel is conduction of water and minerals. It also offers mechanical strength to the plant.

c. Xylem fibres

The fibres of sclerenchyma associated with the xylem are known as xylem fibres or wood fires. They give additional mechanical support to the plant body. They are present both in primary and secondary xylem. Xylem fibres are dead cells and have lignified walls with narrow lumen. Xylem fibres are also called libriform fibres.

d. Xylem (wood) parenchyma

The parenchyma cells associated with the xylem are known as xylem parenchyma.Xylem parenchyma is the only living tissue amongst the constituents of xylem. The cell wall is thin and made up of cellulose. The xylem parenchyma cells store food reserves in the form of starch and fat. They also assist in conduction of water.

Comparison: Vessels and Tracheids

A vessel is made of a number of cells.	A tracheid is formed from a single cell.
The ends are rounded or transverse.	The ends are generally oblique and tapering.
A vessel is several centimeters in length.	A tracheid is only a fraction of a centimeter
The septa between adjacent cells get dissolves to produce a vessel.	The end walls or septa remain intact.
They are comparatively wider.	Tracheids are comparatively narrow.
The wall is less thickened	The wall is more thickened.
The lumen is wide	The lumen is comparatively narrow.

Phloem

Like xylem, phloem is also a complex tissue. It conducts food materials to various parts of the plant. The phloem elements which are formed from the procambium of apical meristem are called primary phloem. The phloem elements which are produced by the vascular cambium are called secondary phloem. The primary phloem elements that develop first from the procambium are smaller in size called the protophloem, whereas those develop later are larger in size called metaphloem. The protophloem is short lived. It is crushed by the developing metaphloem.Phloem is composed of four kinds of cells: sieve elements, companion cells, phloem parenchyma and phloem fibres. Companion cells are present only in angiosperms. Companion cells are absent in pteridophytes and gymnosperms. Phloem fibres are absent in the primary phloem of most of the angiosperms. But they are usually present in the secondary phloem.

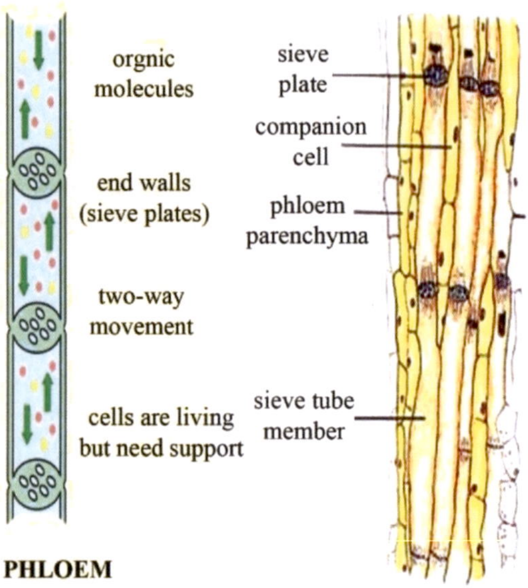

Fig.4.5.2. Phloem and its elements

a. Sieve elements

Sieve elements are the conducting elements of the phloem. They have thick primary walls. Their end walls are transverse or oblique. The end wall contains a number of pores and it looks like a sieve. So it is called a sieve plate.The sieve elements are arranged one above the other and form vertical sieve tubes. In matured sieve tube, nucleus is absent. It contains a lining layer of cytoplasm. This is an important feature of sieve elements. A special protein called slime body is seen in it. The conduction of food material takes place through cytoplasmic strands. They are distinguished into sieve cells and sieve tubes.Sieve cells occur in pteridophytes and gymnosperms, while sieve tubes occur in angiosperms. Sieve cells have sieve areas on their lateral walls only and are not arranged one above the other in linear rows. They are not associated with companion cells.Sieve tubes are arranged one above the other in linear rows and have sieve plates on their end walls. They are associated with the companion cells. In mature sieve elements, sometimes the pores in the sieve plate are blocked by a substance called callus.

b. Companion cells

The thin-walled, elongated, specialized parenchyma cells, which are associated with the sieve elements, are called companion cells. In contrast to sieve elements, the companion cells have cytoplasm and a prominent nucleus. They are connected to the sieve tubes through pits found in the lateral walls. The companion cells are present only in angiosperms and absent in gymnosperms and pteridophytes. They assist the sieve tubes in the conduction of food materials.

c. Phloem parenchyma

The parenchyma cells associated with the phloem are called phloem parenchyma. These are living cells. They store starch and fats. They also contain resins and tannins in some plants. They are present in all, pteridophytes, gymnosperms and dicots. In monocots, usually phloem parenchyma is absent.

d. Phloem or bast fibres

The fibres of sclerenchyma associated with phloem are called phloem fibres or bast fibres. They are narrow, vertically elongated cells with very thick walls and a small lumen (the cell cavity). Among the four kinds of phloem elements, phloem fibres are the only dead tissue.These are the strengthening and supporting cells. These are generally absent in the primary phloem, but occur frequently in the secondary phloem.

Vascular Tissue

Vascular tissue is a complex conducting **tissue**, formed of more than one cell type, found in **vascular** plants. The primary components of **vascular tissue** are the xylem and phloem. These two **tissues** transport fluid and nutrients internally.

5

Anatomy of Stem

Primary Dicot Stem

A stem with only primary permanent tissues is called primary stem. It represents a young stem. The different tissues present in dicot stem are arranged in concentric layers. e.g. Sunflower

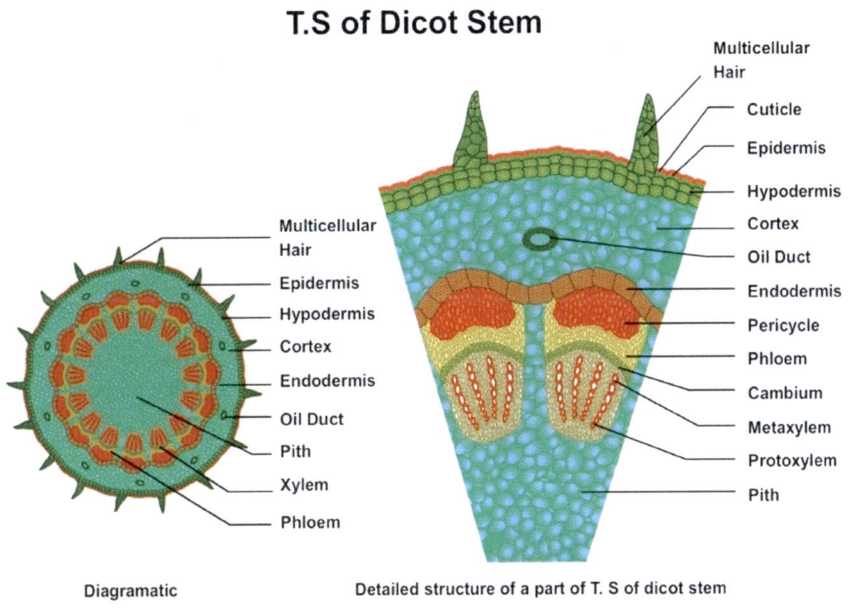

Fig. 5.1: Dicot Stem

I) Epidermis

Epidermis is the outermost layer of the stem. It is made up of compactly arranged parenchymatous cell, which look rectangular in transverse section. The cells are transparent and devoid of chloroplasts. The outer walls are convex, thickened and cutinized. On the outer side they possess a layer of cuticle. The internal walls of the epidermal cells are thin. The radial walls are thick towards the outer side and gradually become thin towards the inner side. Pits occur in the radial walls. The epidermis of sunflower bears several unbranched multicellular hair. At place the epidermis contains minute pores called stomata or stoma. Each stomata or stoma has a pair of specialized kidney shaped cells called guard cells. The guard cells have a few chloroplasts. By their swelling, the two guard cells can form a pore between them.

Functions of epidermis

i. Protection of internal tissues

ii. Prevention of entry of harmful organisms

iii. Minimizing surface transpiration by having thick cuticle

iv. Exchange of gases through the stomata

v. Protection against excessive heating up and sudden changes in temperature with the help of hair

II) Hypodermis

The hypodermis is made of a new layered sub-epidermal collenchyma tissue. Its cell possesses cellulose thickenings either on the tangential walls (lamellate collenchyma) or at the angles (angular collenchyma). The collenchyma cells are green and enclose small inter-cellular spaces.

Functions of hypodermis

i) Providing mechanical strength

ii) Storage of food

iii) Manufacture of food with the help of chloroplast

III) Cortex

It is few to several cells in thickness. The cortex is made up of thin walled angular, oval or rounded parenchymatous cells. They enclose intercellular spaces. The cortical cells may possess chloroplasts. The major function of the cortex is storage of food. Cortex contains a number of longitudinally running oil ducts. Each oil duct has a channel which is lined by an epithelium of small glandular cells.

IV) Endodermis

It is a wavy layer of one cell in thickness. The endodermis lies at the innermost boundary of cortex. It is made up of barrel – shaped cells which do not enclose intercellular spaces. The endodermal cells contain conspicuous starch grains as food reserve. Therefore, the endodermis is also called starch sheath.

V) Pericycle

It is a few layered thick tissues. It lies inner to the endodermis and the outside the vascular strand.The pericycle is made up of both parenchyma and sclerenchyma fibres. Sclerenchyma lies on the outside of the vascular bundles in the form of semicircular patches called bundle caps. Paremchymatous pericycle is present outside medulary rays. The sclerenchymatous pericycle provides mechanical strength to the young tem. The parenchymatous pericycle stores food.

VI) Vascular Strand

The vascular strand is in the form of ring of vascular bundles present around the central pith and inner to the pericycle. The vascular bundles are wedge shapped. Each vascular bundle consists of phloem (primary) on the outside, xylem towards the inside and a strip of cambium in between the two.Phloem and xylem tissues lie on the same radius. Such vascular bundles are known as conjoint (with both phloem and xylem), collateral (phloem and xylem on the same radius) and open (with a strip of cambium in between phloem and xylem)

a) Phloem

It lies towards the pericycle on the outer side of vascular bundle. Phloem consists of sieve tubes, companion cells, phloem parenchyma and some phloem fibres. The companion cells and phloem parenchyma are connected with sieve tubes through pits. They help in the lateral flow of the organic food. The companion cells also control the functions of the sieve tubes. The sieve tubes conduct organic food longitudinally.

b) Xylem

It is found towards the pith or the inner portion of the vascular bundles. Xylem consists of two parts, smaller protoxylem (of narrow elements) and larger metaxylem (of broader elements). Protoxylem or first formed xylem lies at the tip of metaxylem towards the pith or centre of stem. Therefore, the xylem is endarch.**Xylem** consists of tracheids, vessels, xylem parenchyma and xylem fibres. Out of these only the xylem parenchyma cells are living. They are smaller in size than the parenchyma cells found outside the bundles. Most of the xylem parenchyma is found around the protoxylem region. Parenchyma cells store food and help in the lateral conduction of sap. For this, their walls have simple pits.

Vessels are present in the form of a few radial rows. They are angular in outline. The vessels of the protoxylem region are smaller and possess annular or spiral thickenings. These thickenings make the protoxylem vessels elastic and capable of stretching during the elongation of stem. The vessels of the metaxylem have pitted thickenings.

Tracheids are present in between and around the radial rows of vessels especially of the metaxylem region. Xylem fibres lie scattered amongst the tracheids. The vessels, tracheids and xylem fibres, all provide mechanical strength to stem. However, the most important function of xylem is the conduction of water and mineral substances. This is carried out by vessels and tracheids.

Cambium

Cambium is in the form a narrow strip of primary meristematic cells that lies between the phloem and the xylem of a vascular bundle. It is called intrafascicular or fascicular cambium. The cells are thin walled and rectangular in out line. They possess prominent nuclei. It increases the girth of stem by producing secondary phloem towards the outside and secondary xylem towards the inner side.

VII) Medullary or pith rays

They are radial strips of parenchyma which are present between adjacent vascular bundles. The medullary rays connect the pith with pericycle and cortex. The cells are radially elongated. The ray cells make intimate contact with the conducting cells of both phloem and xylem through pits.The medulary rays help in the radial conduction of food and water. They also transport gases from pith to cortex and vice versa.

VIII) Pith or medulla

It forms the centre of the stem. The pith is made up of oval or rounded parenchyma cells which enclose intercellular spaces. The pith cells store food.

Primary Monocot Stem

A monocot stem is lack of secondary growth. It, therefore, possesses only the primary permanent tissues. The various tisues, unlike a dicot stem, are not arranged in concentric rings. The stem can be solid, or fistular. A typical monocot stem consists of the following tissues.

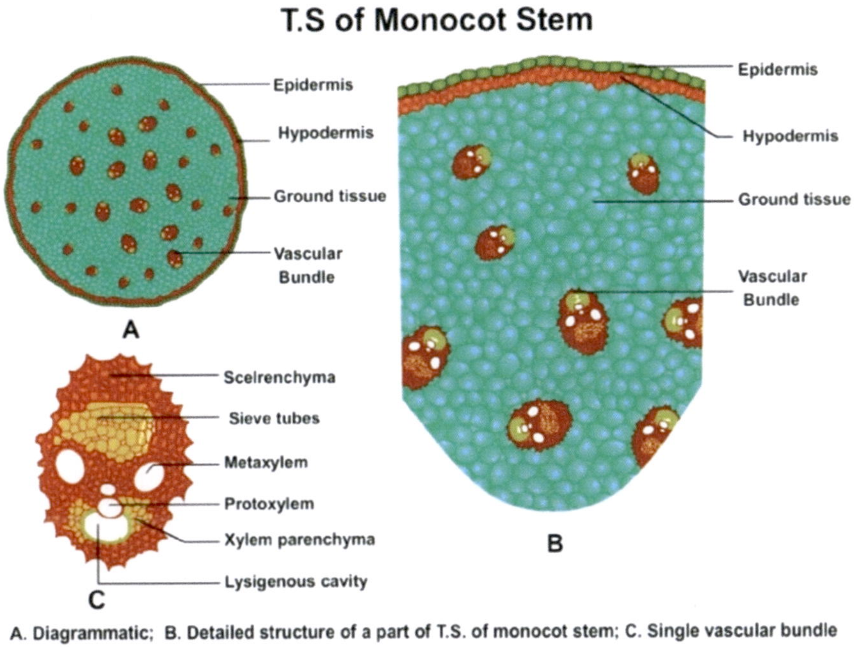

Fig. 5.2: Monocot Stem

Epidermis

It is the outermost layer of the stem which is made up of compactly arranged transparent elongated and rectangular or barrel shaped parenchyma cells. The outer walls of epidermis cells possess deposition of silica and cutin. A separate layer of cuticle also occurs on the outside, the cuticle and cutinized epidermal cells prevent the evaporation of water from the stema. Silica provides stiffness. At places the epidermis possesses stomata for gaseous exchange. Each stoma or stomata has two dumble shaped guard cells. The guard cells can open or close the stoma. They contain a few chloroplasts.

II) Hypodermis

It is 2 – 3 layered thick and lies below the epidermis. The hypodermis is made up of thick walled sclerenchyma fibres. It acts as a heat screen and provides the rigidity and mechanical strength to stem.

III) Ground Tissue

It is parenchymatous and occupies the whole stem interior. In the stem of maize the cells are small and angular towards the hypodermis but become large and oval in the inner region. The ground tissue stores food. Some of the outer cells may also synthesize food due to the presence of chloroplasts in them. Abundant inter-cellular spaces are present in the ground tissue. These spaces communicate with exterior through the stomata present in the epidermis. In some monocots a central cavity is formed in the ground tissue by the disintegration of cells during growth (e.g. grass). A band or cylinder of sclerenchyma occurs in the ground tissue of grass stem.

IV) Vascular Strand

The vascular strand is in the form of large number of vascular bundles. They lie scattered throughout the ground tissue (atactostele). Vascular bundles are smaller but more numerous towards the outside than towards the centre. The vascular bundles are oval or round in outline. They contain both phloem and xylem. Phloem lies towards the outside and the xylem on the inner side. Cambium is absent. The vascular bundles are, therefore, conjoint, collateral but closed. Each vascular bundle is surrounded by a sheath of sclerenchyma known as bundle sheath.The monocots sheath is more developed on outer and the inner sides. The hypodermis and bundle sheath coalesce in some of the outer vascular bundles. The bundle sheath is absent in ***Asparagus.***

Phloem consists of sieve tubes, companion cells and a few phloem fibres. Phloem parenchyma is absent.The sieve tubes conducts organic food. In maize, the phloem is distinguished into outer protophloem and inner metaphloem. The protophloem gets crushed in the latter stages.Xylem is in the form of letter Y. It is endarch, i.e., protoxylem lies towards the centre of the stem. Xylem is made up of vessels, tracheids, xylem parenchyma and a few xylem fibres. Metaxylem generally consists of two large oval or rounded vessels is lying at the upper two angles of xylem. The metaxylem vessels have pitted walls. The two vessels are connected with each other by polygonal tracheids having pitted thickenings. Protoxylem has a few small oval vessels. They lie at the lower angle of xylem. The vessels of the protoxylem show spiral and annular thickenings. Xylem parenchyma and a few fibres are found just outside them.Some of the protoxylem vessels and xylem parenchyma cells dissolved or separate during the rapid growth of the stem to form a cavity of maize are schizo – lysigenous in origin. It generally stores water. The tracheids and vessels help in the conduction of sap as well as mechanical support. In maize the protoxylem cavity and protophloem can be absent in the smaller vascular bundles.

Comparison: Dicot and Monocot Stem

S.No.	Dicot Stem	Monocot Stem
1	Stomata have kidney –shaped guard cells.	Stomata usually possess dumbel shaped guard cells.
2	The hypodermis is made up of collenchymas which may be green.	The hypodermis is formed of non green sclerenchyma fibres.
3	The internal tissues are arranged in concentric layers.	The concentric arrangement of tissues is absent.
4	The ground tissue is differentiated into cortex, endodermis, pericycle, pith, etc.	The ground tissue is a mass of similar cells.
5	The stem is almost always solid.	The stem is generally hollow in the centre (exception in maize).
6	The vascular bundles are arranged in ring around the pith.	The vascular bundles are scattered throughout the ground tissue.
7	The vascular bundles are few in number and are of similar size	The vascular bundles are numerous and are of different size –smaller towards the outside and larger towards the centre
8	The vascular bundles are wedge shaped in outline.	They are oval or rounded in outline.
9	No bundle sheath is present on the outside of a vascular bundle.	A sclerenchymatous bundle sheath is generally present on the outside of each vascular bundle.
10	The vascular bundles are open due to the presence of cambium in between phloem and xylem.	The vascular bundles are closed.
11	The stem shows secondary growth due to the formation of secondary vascular tissues and periderm.	Secondary growth is usually absent.
12	Phloem parenchyma is present in the phloem along with other elements.	Phloem parenchyma is absent.
13	The vessels are polygonal in outline.	The vessels are oval or rounded.
14	The vessels are usually arranged in chains or rows	The vessels are arranged in a Y- shapped manner.
15	Metaxylem vessels are generally numerous.	Metaxylem vessels are few in number.
16	Medullary rays are found in between the vascular bundles	Medullary rays are absent.
17	A cavity is not found in the vascular bundle.	A cavity containing water is found in vascular bundle be the dissolution or separation of some protoxylem vessels and parenchyma lying nearby.
18	The older vascular tissues stop functioning after some time. They are replaces by younger vascular tissues.	The first formed vascular tissues continue functioning throughout the life of the plant.

6

Anatomy of Root

Primary Dicot Root

A young dicot root which does not possess secondary tissues but only the primary permanent tissues (tissues derived from the apical growing point) is called primary dicot root. It is generally cylindrical in outline and possesses the following tissues.

I) Epiblema or Piliferous Layer

It is the outermost layer of the root. It is made of compactly arranged thin walled and slightly elongated parenchymatous cells. Epiblema of a root differs from the epidermis of the stem in being devoid of distinct cuticle and stomata. Some cells of the epiblema give rise to thin – walled tubular outgrowths called root hairs. In some plants the root hair mother cells are shorter in comparison to other epiblema cells. The root hairs and piliferous layer (pilus – hair, ferre to carry). The root hair and thin walled epiblema cells absorb water and mineral salts from the soil. Root hairs commonly do not live for more than one week. With their death the epiblema cells become suberised and cutinised.

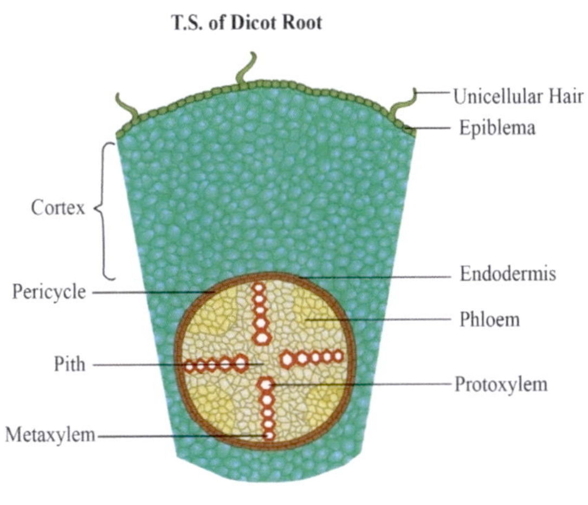

Fig. 6.1: Dicot Root

II) Cortex

It lies below the epiblema. The cortex is made up of many layers of thin walled parenchyma cells. The parenchyma cells may be rounded or angular. They enclose inter-cellular spaces for diffusion of gases.The cells of the cortex store food. They also conduct water from the epiblema to the inner tissues.

III) Endodermis

It is made up of single layer of barrel – shaped cells which do not enclose inter-cellular spaces. The cells are rich in starch grains.The young endodermal cells possess a band of thickening which runs along their radial and tangential walls. This band of thickening is called casparian strip. It is made up of both suberin and lignin. In a transverse section, the casparian strip appears in the form of small lenticular swellings on the radial walls only. The casparian strip may become faint due to deposition of additional wall thickening of lignin. In such cases small unthickened areas called pits are left in the cell walls.The endodermis probably controls the movement of fluid and air from outside to the interior and vice versa.

IV) Pericycle

Endodermis is followed by one or more layers of pericycle. The cell of the pericycle are thin walled and parenchymatous. The pericycle is a very important layer. A part of the vascular cambium is formed by the pericycle. The cork cambium also develops from it. All lateral roots originate from the pericycle.

V) Vascular Strand

Inner to the pericycle are found a few alternately arranged bundles of xylem and phloem. They are equal in number and lie on different radii. Such vascular bundles are called radial bundles.The various xylem bundles put together give a stellate or star shaped appearance. The number of rays is equivalent to the number of xylem bundles (and phloem bundles). According to the number of rays, the root may be diarch (with 2 Xylem bundles e.g. Tomato), triarch (Pea), tetrarch (Sunflower), pentarch (with 5 xylem bundles) or polyarch (with more than 5 xylem bundles). Protoxylem or the first formed xylem lies at the tip of the rays while metaxylem or later formed xylem is present towards the centre of the root. Such a xylem is called exarch (ex- outside, arche – beginning). The metaxylem of different xylem bundles may lie separate from one another so that a pith is present in the centre of the root (e.g. Bean). However, usually the metaxylem elements of different bundles meet in the centre to form a solid star shaped structure. In such case the pith is absent.

Xylem is made up of vessels and few tracheids. Vessels and tracheids are polygonal in outline. Protoxylem elements are fewer, smaller and narrower. The metaxylem elements are larger and wider. They have pitted thickenings while protoxylem possesss spiral, annular, reticulate or scalariform thickenings.

Xylem performs two important functions

- Mechanical strength
- Conduction of water and mineral salts to the shoot

In between the two adjacent xylem bundles is found a phloem bundle. Phloem and xylem bundles are separated from each other by one or more layers of small thin walled cells called conjunctive parenchyma or tissue. Latter on the conjunctive tissue becomes meristematic to form vascular cambium. In some roots the phloem bundle bear a cap of sclerenchyma towards the outer side, below the pericycle.**Phloem** consists of sieve tubes, companion cells and phloem parenchyma. It conducts organic food from shoot to the root and its branches.

Primary Monocot Root

There is no distinction between a young and an old root of monocotyledonous plant. This is due to the absence of secondary growth in monocot roots. Typical monocot root consists of the following parts.

I) Epiblema or piliferous layer

It is the outermost layer of young root which has thin walled cells. Some of the cells give rise to root hairs.Root hairs are tubular in outline and lie in contact with soil water. Both epiblema and root hairs are devoid of cuticle. They take part in the absorption of water and mineral salts. In older parts the epiblema is shed or become impervious.

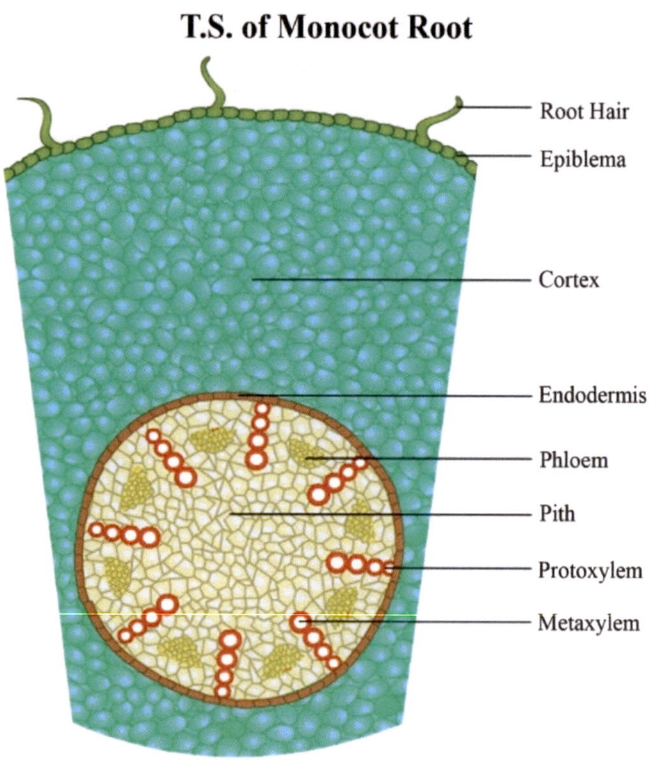

Fig. 6.2: Monocot Root

II) Cortex

It is very wide region of parenchymatous cells that enclose intercellular spaces for the exchange of gases. The cells are store food. In older roots the outer one or more layers of the cortex become thick walled and suberised. They constitute the exodermis. It is protective and to some extent absorptive in function.

The cortex of a monocot root has, therefore, three functions

- Storage of food
- Conduction of water from the root hairs to the inner tissues
- The outermost layer or layers of the cortex produce protective exodermis in the older roots.

III) Endodermis

It is a single layered structure that lies on the inner boundary of the cortex. It is made up of barrel-shaped cells which do not enclose intercellular spaces. The young endodermal cell possesses an internal strip of suberin and lignin which is known as casparian strip. However, it soon disappears due to the additional thickening of the endodermal cells. Endodermal cells lying opposite the protoxylem groups, however, remain in the primary stage with the usual casparian strip. These unthickened cells are called passage cells. The passage cells are ment for the conduction of fluids inwardly from the cortex and outwardly from the interior into the cortex. The thickened cells can also allow transport through plasmodesmata of pits.The endodermis regulates the flow of fluid both inwardly as well as outwardly.

IV) Pericycle

It lies below the endodermis. The pericycle is made up of parenchymatous cells which store abundant food. In monocots the pericycle does not form cambia. It, however, produces lateral roots.

V) Vascular strand

It is in the form of several alternate and radial xylem and phloem bundles. The number is 100 or more of each type in *Pandanus* and Palms. In many cases the vascular bundles are embedded in a cylinder of sclerenchyma. The vascular bundles are arranged in the form of ring around a central pith. The xylem bundles are exarch, i.e., protoxylem lies towards the outside while the metaxylem faces inwards.Xylem is made up of rounded or oval vessels and xylem parenchyma. Xylem provides mechanical strength and helps in the conduction of water and mineral salts.Phloem bundles alternate with the xylem bundles. These two are separated from each other by means of narrow strip of parenchyma called conjunctive tissue. The cells of this tissue simply store food. They do not take part in the formation of cambium.Phloem consists of sieve tubes, companion cells and phloem parenchyma, it helps in the conduction of organic food.

VI) Pith

The centre of monocot root is occupied by the pith. It consists of parenchymatous cells which may be rounded or angular. Intercellular spaces are present amongst the pith cells. The pith cells store food.

Comparison: Dicot and Monocot Root

S.No.	Dicot Root	Monocot Root
1	Cortex is comparatively narrow	Cortex wider
2	The epiblema, the cortex and even the endodermis are peeled off and replaces by cork.	Cork is not formed. The cortex and the endodermis persist. Only the epiblema is peeled off.
3	Endodermis is less thickened and casparian strips are more prominent.	Casparian strips are visible only in young root. The endodermal cells later become highly thickened
4	Passage cells are generally absent in endodermis.	Thin walled passage cells generally occur in the endodermis opposite the protoxylem points.
5	Pericycle produces lateral roots cork cambium and part of the vascular cambium.	Pericycle produces lateral roots only.
6	The number of xylem and phloem bundles varies from 2- 5 or some times 8.	Xylem and phloem bundles are numerous and are 8 or more in number.
7	Xylem vessels are generally angular.	Xylem vessels are oval or rounded.
8	Conjunctive tissue is parenchymatous.	Conjunctive tissue may parenchymatous or sclerenchymatous.
9	Conjunctive parenchyma forms the cambium.	Conjunctive parenchyma does not produce cambium.
10	Secondary growth takes place with the help of vascular and cork cambium.	Secondary growth is absent.
11	Pith is either absent or very small.	A well – developed pith is present in the centre of the root.

7

Secondary Growth

Secondary Growth in Dicot Stem

The formation of secondary tissue from the lateral meristem is called secondary growth. It increase in diameter of the stem. A stem with secondary tissues is called secondary stem. Secondary tissues are formed by two types of lateral meristem, vascular cambium and cork cambium or phellogen. Vascular cambium produces secondary vascular tissues while phellogen form periderm. Secondary growth occurs in perennial dicots such as trees and shrubs. It is also found in the woody stems of herbs. In such cases, the secondary growth is equivalent to one annual ring, e.g. sunflower.

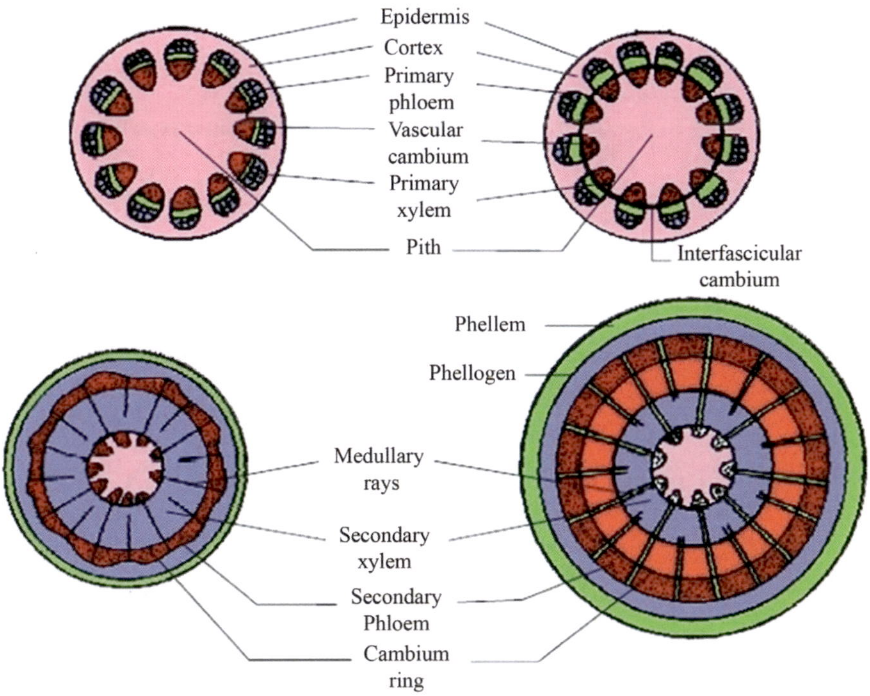

Fig. 7.1: Secondary growth in a dicot stem (diagrammatic)- stages in transverse views

Formation of secondary vascular tissues

They are formed by the vascular cambium. Vascular cambium is produces by two types of meristems: **intrafascicular** and **interfascicular cambium**.

1. Intrafascicular cambium is a primary meristem which occurs as strips in vascular bundles.
2. Interfascicular cambium arises secondarily from the cells of medullary rays which occur at the level of intrafascicular strips.
3. These two types of meristematic tissues get connected to form a ring of **vascular cambium**

The cells of vascular cambium are of two types: **fusiform initials** and **ray initials**.

1. The ray initials are smaller and are found in groups they give rise to vascular rays.
2. The fusiform initials are elongated and taper at their ends.

They divide both on the outer and inner sides to form tissue mother cells. The tissue mother cells, formed outside the ring of vascular cambium, give rise to secondary phloem while those of the inner side produce secondary xylem. With the formation of secondary xylem on the inner side, the vascular cambium moves gradually to the outside by adding new cells. The phenomenon is called dilatation. New ray cells form additional rays every year.

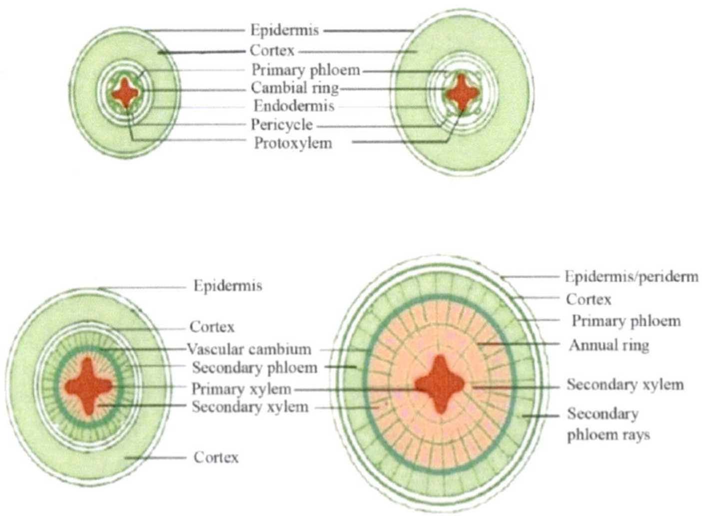

Fig. 7.2: Different stages of the secondary growth in a dicot root

a) Vascular rays

The vascular rays or secondary medullary rays are rows of radially arranged cells which are formed in the secondary vascular tissues. They are a few cells in height.Depending upon their breadth, the vascular rays are uniseriate (one cell breadth) or multiseriate (two or ore cells in breadth).Vascular rays may be homocellular (having one type of cells) or heterocellular (with more than one type of cells). These cells of the vascular rays enclose intercellular spaces. The part of the vascular ray present in the secondary xylem is called wood or xylem ray while the part present in the secondary phloem is known as phloem ray.The vascular rays conduct water and organic food and permit diffusion of gases in the radial direction. Besides, their cells store food.

b) Secondary phloem

The secondary phloem forms a ring on the outside of cambium. It is made up of the same types of cells as are found in the primary phloem. Fibres commonly occur either in patches or bands. The primary or older phloem present on the outside gets crushed.

c) Secondary xylem

It forms the bulk of the stem and is commonly called wood. Its width grows with the age of plant.The primary xylem persists as conical projection on its inner side. The yearly growth of secondary xylem is distinct in the ares which experience two seasons, one favourable (spring or rainy season) and the other unfavourable (autumn, winter or summer).Here the annual or yearly growth appears in the form of distinct rings which are called annual rings.

Annual ring

It is the wood formed in a single year. It consists of two types of wood, spring wood and autumn wood.The spring or early wood is much wider than the autumn wood. It consists of large xylem elements. The autumn or late wood contains compactly arranged smaller elements which have comparatively thicker walls. In autumn wood, tracheids and fibers are more abundant than those found in the spring wood.The transition from spring to autumn wood in an annual ring is gradual but the transition from autumn wood to the spring wood of the next year is sudden. Therefore, each year's growth is quite distinct. The number of annual rings corresponds to the age of that part of the stem. They can be counted by increment borer. Besides giving the age of the plant, the annual rings also give some clue about the climatic conditions of the past through which the plant has passed. Ring like markings occur in wood of some plants growing in uniform climate, e.g., Mango. They represent tangential bands of xylem parenchyma.

Sapwood and heartwood

The wood of the older stems gets differentiated into two zones, the outer lighter and functional sapwood or alburnum and the inner darker and nonfunctional heartwood or duramen. The tracheids and vessels of the heartwood get plugged by the growth of the adjacent parenchyma cells into their cavities through the pits. These ingrowths are called tyloses. Ultimately, the parenchyma cells become lignified and dead. Various types of plant products like oils, resins, gums and tannins are deposited in the cells of the heartwood. These substances provide colour to the heartwood. They are also antiseptic. The heartwood is, therefore, stronger and more durable than sapwood. It is however, liable to be attacked by wood rotting fungi. Hollow tree trunks are due to their activity.

Formation of periderm

In order to provide for increase in girth and prevent harm on the rupturing of the outer ground tissue due to the formation of secondary vascular tissues, dicot stems produce a cork cambium phellogen in the outer cortical cells. Rarely it may arise from the epidermis (e.g. Teak, Olender), hypodermis (e.g. Pear) or phloem parenchyma. Phellogen cells divide on both the outer as well as the inner side (Bipolar) to form secondary tissues. The secondary tissue produced on the inner side of the phellogen is parenchymatous or collenchymatous. It is called secondary cortex or phelloderm. Its cells show radial arrangement. Phellogen produces cork or phellogen on the outer side. It consists of dead and compactly arranged cells that possess lignified and suberised cell walls. The cork cells contain tannins. Hence they appear brown or dark brown in colour.The cork cells of some plants are filled with air (e.g. Bottle Cork). The phelloderm, phellogen and phellem together constitute the periderm.Cork prevent the loss of water by evaporation. It also protect the interior against entry of harmful micro-organisms, mechanical injury and extremes of temperature. At places phellogen produces aerating pores instead of cork. These pores are called lenticels. Each lanticel is filled by mass of somewhat loosely arranged subrised cells called complementary cells.

Bark

All the dead cells lying outside phellogen are collectively called bark. The outer layers of the bark are being constantly peeled off on account of the formation of new secondary vascular tissues in the interior. The peeling of the bark may occur in sheets (sheets or ring bark, e.g. *Eucalyptus*) or in irregular strips (scaly bark). The scaly bark is formed when phellogen arise in strips instead of ring, e.g., *Acacia* Bark is insect repellent, decay proof, fire – proof and acts as a heat screen.Commercially it is employed in tanning (e.g., *Acacia*), drugs (e.g., *Cinchona quinine*) or as spice (e.g., *Cinnamon*). The cork of *Quercus suber* is employed in the manufacture of bottle stoppers, insulators, flots, sound proofing and linoleum

Lenticels

Lenticels are aerating pores in the bark of plants. They appear on the surface of the bark as raised scars containing oval, rounded or oblong depressions.A lenticel is commonly produced beneath a former stomata or stoma of the epidermis. Its margin is raised and is formed by surrounding cork cells. The lenticels is filled up by loosely or compactly arranged thin walled rounded and suberised or unsuberised cells called complementary cells. They enclose intercellular spaces for gaseous exchange. The complementary cells are formed from loosely arranged phellogen cell and division of sub-stomatal parenchyma cells. The suberised nature of complementary cells checks excessive evaporation of water. In temperate plants the lenticels get closed during winter by the formation of compactly arranged closing cells over the complementary cells.

Comparison: Primary and Secondary Xylem

Sr.No.	Primary xylem	Secondary xylem
1	It is formed from apical meristem.	Secondary xylem is produced from a lateral meristem.
2	Primary xylem occurs in all types of stems and root.	It is found only in dicots with the exception of some annuals.
3	The xylem is differentiated into two parts, protoxylem and metaxylem.	There is no distinction.
4	A radial system is absent.	It is traversed by a radial system of xylem rays
5	Annual rings are absent.	It may show annual rings.
6	There is no distinction into sapwood and heartwood.	A distinction into sapwood and heartwood is found in large woody plants.
7	Fibres are few or absent.	Fibres are generally abundant.
8	The tracheids and vessels are long and comparatively thin walled.	The tracheids and vessels are long and comparatively shorter and more thick walled.
9	Tyloses are absent.	The tracheids and vessels of older xylem get blocked by the development of tyloses.

Secondary growth in Dicot Root

Secondary growth is the production of secondary tissues. Secondary tissues increase the thickness of the root. With the exception of some annuals, most of the dicots show secondary growth in their roots. It takes place by the production of two types of secondary tissues. They are secondary vascular tissue and epiderm. These tissues are formed by meristems, vascular cambium and cork cambium respectively.

Secondary vascular tissues

The conjunctive parenchyma cells lying around each phloem bundle become meristematic to produce a strip of cambium. These strips are soon joined by cambial strips arising from pericycle cells lying opposite the protoxylem points. Wavy band of vascular cambium is, therefore, formed the cambium of the root is a secondary meristem.The vascular cambium becomes active first below the phloem bundles. The vascular cells formed by it push the primary phloem to outside. The cambium also moves outwardly and becomes circular in outline. The ring of vascular cambium produces secondary xylem on the inner side and phloem to the outside. Both of them are in the form of rings (primary vascular bundles). The primary phloem gets crushed by the growth of secondary vascular tissues.The older secondary phloem is also partially destroyed as the new phloem becomes functional. The primary and secondary xylems persist. Primary xylem is distinguishable by its exarch nature and central position. As compared to the primary xylem, the vessels of the secondary xylem are broader and thinner.

Annual rings are not very sharp because unlike aerial climate, the climate of the soil does not vary much during different seasons. The secondary phloem is made up of sieve tubes, companion cells and phloem parenchyma. Sclerenchyma fibres are rare. The secondary xylem is formed of vessels trachieds and xylem parenchyma. At place, the vascular cambium possesses ray initials. They produce vascular rays. The rays are made up of two parts, xylem or wood ray and phloem ray. They help in radial conduction of substances. Very prominent vascular rays are produces opposite the protoxylem point. They are many cells in which in width (multiseriate)

Secondary ground tissues or Periderm

The pericycle layer either directly or after a few divisions, becomes converted into a secondary meristem called cork cambium or phellogen. Rarely phellogen appears in the cortex. The cells of phellogen divide both towards, the outside as well as inside. The tissue formed towards the inner side is parenchymatous and known as secondary cortes or phelloderm. It is only a few layers in thickness. The cells formed on the outside by the phellogen become dead. The cavities get filled up with tannin and their walls become suberised. They are described as cork cell. The tissue of cork cell is spoken as cork or phellem. The cork is impervious to water. It protecy the interior from mechanical injury and entry of bacteria.Primary tissues present outside the cork undergo starvation and get shriveled. Under the impact of secondary growth in the interior, the outer layers of the cork are also peeled off occasionally.

The phellem the phellogen and the phelloderm are collectively called secondary ground tissue or periderm. At places the phellem or cork bears lenticels for exchange of gases.

Origin of lateral roots: The root branches are endogenous (endon within, genes – produces) in origin. It is aprotective mechanism. The lateral roots arise from pericycle (endodermis in case of seedless vascular plants) opposite the protoxylem points, a little distance behind the tip.Two or more cells of the pericycle become meristematic. They develop into the growing point of the lateral root. This soon produces root cap, protoderm, ground meristem and procambium. Latter these give rise to epiblems, ground tissue and vascular strand respectively. The vascular strand of lateral root gets connected with that of the parent root. Some of the cells present at the tip of the lateral root may act as digestive cells for destroying the outlying cells. The endodermis, the cortex and the epiblem of the parent root ultimately break and the lateral root comes out.

8

Tissue System

The tissue system

A group of tissues performing a similar function irrespective of its position in the plant body is called a tissue system. In 1875, Sachs recognized three tissue systems in the plants.

a) Epidermal tissue system

b) Vascular tissue system

c) Fundamental tissue system.

a) Epidermal tissue system

Epidermal tissue system is the outermost covering of plants. It consists of epidermis, stomata and epidermal outgrowths. Epidermis is generally composed of single layer of parenchymatous cells compactly arranged without intercellular spaces. But it is interrupted by stomata. In leaves, some specialized cells which surround the stomata are called the guard cells. Chloroplasts are present only in the guard cells of the epidermis. Other epidermal cells usually do not have chloroplasts. The outer wall of epidermis is usually covered by cuticle. Stoma is a minute pore surrounded by two guard cells. The stomata occur mainly in the epidermis of leaves. In some plants such as sugarcane, the guard cells are bounded by some special cells. They are distinct from other epidermal cells. These cells are called subsidiary or accessory cells. Trichomes and root hairs are some epidermal outgrowths. The unicellular or multicellular appendages that originate from the epidermal cells are called trichomes. Trichomes may be branched or unbranched. Rhizodermis has two types of epidermal cells - long cells and short cells. The short cells are called trichoblasts. Root hairs are produced from these trichoblasts.

Functions of epidermal tissue system

1. This tissue system in the shoot checks excessive loss of water due to the presence of cuticle.

2. Epidermis protects the underlying tissues.
3. Stomata involve in transpiration and gaseous exchange.
4. Trichomes are also helpful in the dispersal of seeds and fruits.
5. Root hairs absorb water and mineral salts from the soil.

b) Vascular tissue system

The vascular tissue system consists of xylem and phloem. The elements of xylem and phloem are always organized in groups. They are called vascular bundles. In dicot stem, the vascular bundle consists of cambial tissue in between xylem and phloem. Such vascular bundle is called **open vascular bundle.** In monocot stem, cambium is absent in the vascular bundle, hence it is known as **closed vascular bundle** In roots, xylem and phloem are arranged in an alternate manner on different radii. It is called **radial arrangement**.In stems and leaves, xylem and phloem are arranged at the same radius and form a vascular bundle together. Such vascular bundle is called conjoint vascular bundle. Depending upon the mutual relationship of xylem and phloem, conjoint vascular bundles are divided into three types. They are **collateral, bicollateral and concentric.**

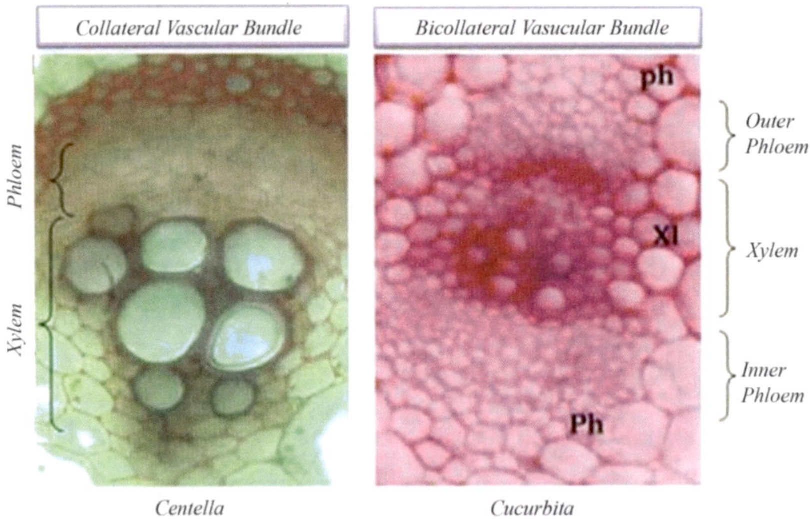

Fig. 8.1. Vascular Tissue

If xylem and phloem in a vascular bundle are arranged along the same radius with phloem towards the outside, such vascular bundle is called collateral vascular bundle.

Bicollateral vascular bundles

If phloem occurs on both the outer and inner sides of xylem, the bundle is called bicollateral. Bicollateral vascular bundles are most typically seen in Cucurbitaceae.

Concentric vascular bundle

The bundle in which either phloem surrounds the xylem or xylem surrounds the phloem completely is known as concentric vascular bundle.

This is of two types **amphicribral** and **amphivasal.** In **amphicribral** concentric vascular bundles, the phloem completely surrounds the xylem. eg. Polypodium.

In **amphivasal** concentric vascular bundles, the xylem completely surrounds the phloem. eg. *Acorus*.

In roots, protoxylem vessels are present towards the periphery and the metaxylem vessels towards the centre. This arrangement of xylem is called exarch.

In stem, protoxylem vessels are towards the centre, while metaxylem towards the periphery. This condition is known as endarch.

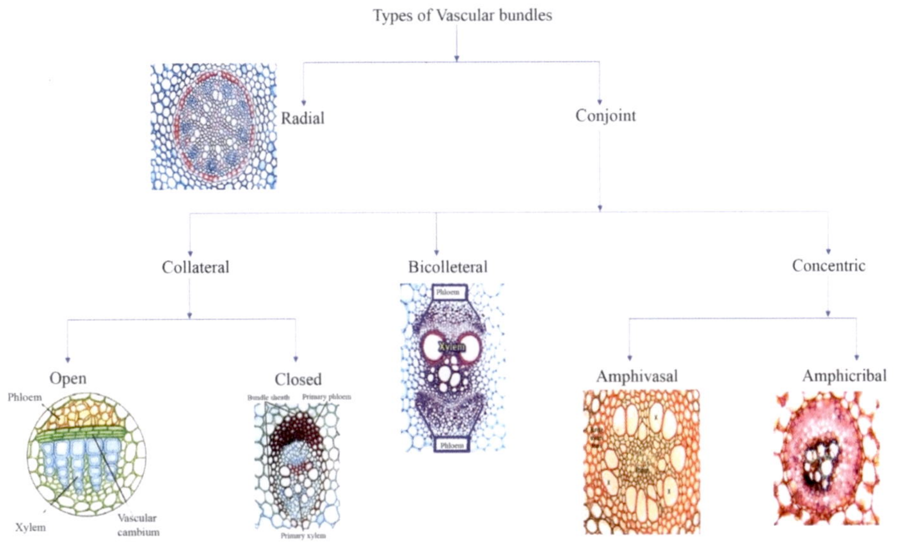

Fig. 8.2: Different types of vasular bundles

c. Ground or fundamental tissue system

The ground or fundamental tissue system constitutes the main body of the plants. It includes all the tissues except epidermis and vascular bundles. In monocot stem, ground tissue system is a continuous mass of parenchymatous tissue in which vascular bundles are found scattered. Here ground tissue is not differentiated into **cortex, endodermis, pericycle and pith**. Generally in dicot stem, ground tissue system is differentiated into three main zones - **cortex, pericycle and pith**.The cortex occurs between the epidermis and pericycle. Cortex may be a few to many layers in thickness. In most cases, cortex is made up of parenchyma tissues. Intercellular spaces may or may not be present. Cortical cells may contain non-living inclusions like starch grains, oils, tannins and crystal. In the leaves, the ground tissue consists of chlorenchyma tissues. This region is called mesophyll. The inner most layer of the cortex is called endodermis. Generally endodermis is made up of barrel shaped parenchyma cells. These cells are arranged in a single layer without intercellular spaces.Pericycle occurs between the endodermis and the vascular bundles. It is generally made up of parenchyma cells. Lateral roots originate from the pericycle. Thus their origin is endogenous. The central part of the ground tissue is known as pith or medulla. Generally this is made up of thin walled parenchyma cells which may be with or without intercellular spaces. The cells in the pith generally store starch, fatty substances, tannins, phenols, calciumoxalate crystals, etc.

The mechanism of wood formation

The growth of the tree is the result of production of new cells by tissues termed meristems.Meristems consist of cells that are undifferentiated and retain the ability to divide and produce new cells. The increase in the height of tree stem is designated as primary growth and is the result of cell production by apical meristems located at the tip of the stem. Just below the apical meristems, some of the cells form a lateral meristem called the cambium. Secondary growth (to diameter), begins with formation of the cambium.Longitudial cells of wood are produced by cambial cells designed as fusiform initials and transverse by ray initials. Cambium initials divide periclinally, namely with walls parallel to the cambium layer and produce xylem and phloem mother cells. The phases of development of wood cells are: cell division, cell enlargement, cell wall thickening, lignification and death.The first phase, cell division, occurs when a cambial initial divides, forming two cells.The outermost cell remains meristematis and the innermost cell develops into a mature wood cell. During the enlargement phase, the cell grows in length and diameter. Softwood cells increases only in radial direction.During the enlargement phase, a very thin and plastic cell wall (primary wall) encases the protoplasm. During next phase, wall thickness is increased by the addition of secondary wall lignification involves the formation of lignin between the newly formed cells and within their cell walls.For most wood cells, death occurs immediately after lignification. However, for those wood cells that perform the function of storage, this step is postponed for an indefinite period.Cells performing the support and conduction roles usually pass through the five phases of cell development in 14 to 21 days.

Formation of Earlywood and Latewood

The transition from earlywood to latewood formation is a conspicuous developmental in trees. Latewood is induced during the later part of the growing season, when cell division activity in the cambial meristem declines. It involves a reduction in radial expansion and an increase in wall thickening of the cambial derivatives. Thus, earlywood is characterized by large-diameter and thin-walled tracheids/vessels, whereas latewood is composed of narrow diameter tracheids/vessels with thick cell walls. The process of earlywood and latewood formation, that is the frequency of division of cambial cells, the enlargement of newly formed xylem cells, and cell wall thicken are controlled by photosynthate activity and plant hormones. In active season IAA **(Indole-acetic acid)** is largely synthesized by buds in the tree crown, developing leaves shoots and in the dormant period in winter, IAA is synthesized by dormant buds and old leaves.The occurrence of small amounts of endogenous IAA in the cambial zone suggests the formation of IAA by cambial cells.IAA synthesized by the tree crown is basipetally transported mainly to cambial cells and their differentiating cells. Concentration of endogenous IAA in the cambial zone of a tree stem exhibits a seasonal change. The concentration of IAA increased from spring to early summer and then decreased toward the autumn to the level found in the spring.In winter, IAA was present at a low level. The stage of rapid decrease of IAA coincides well with the change from earlywood to latewood formation.

IAA is a major factor in the control of elongation of xylem differentiating cells, the formation of earlywood and latewood and the transition of earlywood to latewood.Even if IAA is supplied, cambial activity ceased at a certain time of the season. For termination of cambial activity (dormancy), other factors may be involved.

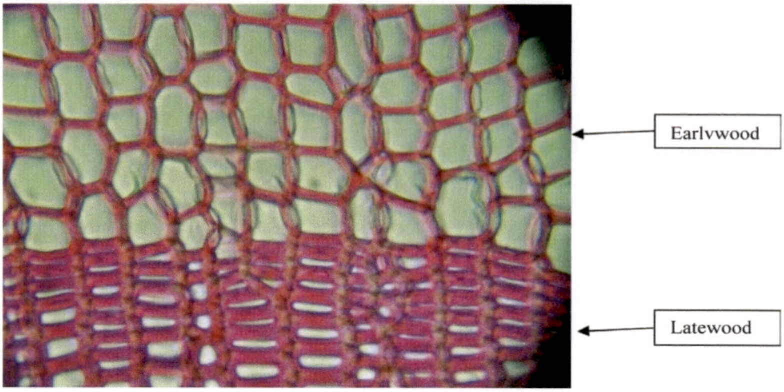

Fig. 8.3: Early and Late wood

A change in the sensitivity of cambial cells to IAA has been suggested to be one of the factors involved. When IAA was supplied to tree stems cambial activity increased, but the effect changed seasonally and after summer the effect was very small.The extract natural of any cambial sensitivity to IAA is scarcely known, but it seems that the change is induced by a change of temperature and day length and is related to structural and histochemical changes in the cambial cells. It was found that a reconstructure of cell membrane occurred before dormancy. If IAA receptor or carrier proteins are present on the plasma membranes these proteins may tentatively be transported of decreased, thus inducing a low sensitivity for IAA. Other hormones like Abscisic Acid (ABA) and cytokinins also play a role in the formation of early and late wood.Other factors responsible for the formation of early and late wood are abundance of soil moisture nutrients.Water deficit conditions induce the early formation of the late wood. But, continued moisture deficit are reported to shorter period of latewood formation.

Comparison: Early Wood and Late Wood

S.No.	Early wood or Spring wood	Late wood or Autumn wood
1.	Formed during spring season.	Formed during winter season
2.	Formed early in a year.	Formed after the early wood.
3.	Consists of xylem tissues with wider vessels.	Consists of xylem elements with narrow vessels.
4.	Produced more in amount.	Produced less in amount.
5.	Less dense.	More dense.
6.	A broad zone of wood.	A narrow zone of wood.
7.	Not as strong as late wood.	Stronger than early wood due to larger Volume of wall materials.

Growth Ring Formation

The activity of vascular cambium is markedly effected by variation in the climate. Growth rings are the result of new growth in the vascular cambium, a layer of cells near the bark that is classified as a lateral meristem; this growth in diameter is known as secondary growth. Visible rings result from the change in growth speed through the seasons of the year; one ring generally marks the passage of one year in the life of the tree.During winter, the cambium stop dividing. In spring season it regains its activity and starts dividing. During the season vegetative growth of the trees is pronounced, more leaves appear and there is dire need of effective transport of water and minerals dissolved in it and food material so that plant can cope with its increasing requirements. The cambium becomes more active because of favourable environments and divides activity producing a large number of vessels that have wider lumens. The secondary xylem formed during this period of pronounced activity is called spring wood. During the autumn season the activity of the cambium is lessened and the vessels produced are generally of smaller size and have narrow limens. Tracheids and wood fibres are produced in large numbers.

The secondary wood formed during this season of the year is called autumn wood or summer wood. In transverse section of the stem these two types of woods appear in the form of distinct concentric circles known as the annual or growth rings.Spring wood circle and autumn wood circle constitute annual rings. Like this, year after year such rings appear and their identity is well marked.The number of annual rings in the old parts of the tree corresponds to its age.In some woody trees e.g. Tilia, Black Ash and *Tamarix dioca*, the vessels in the springs woods are large and arranged in a ring and narrow vessels of the summer or autumn wood are scattered. Such a wood said to be ring porous. In *Eugenia* and *Azadiracta indica* the vessels are more or less uniformly distributed through the spring wood and autumn wood such wood is called diffuse porous wood. The growth rings are not well marked in the trees growing in the tropical region of the world. Here the number of growing rings does not correspond to the age of the tree.The thickness of the annual rings is greatly influenced by fluctuation in the environmental conditions.It has been found that the annual rings remain thin during low rainfall and became thick if the rainfall is heavy.Sometimes permanent changes are brought about in he environment by chopping down the trees or by draining away water. In such cases the thickness of the ring produced before and after the change is remarkable and it has been possible to exactly judge the year when such a change was brought about.

Ray initial and fusiform initials, anticlinal and periclinal division

The vascular cambium produces secondary xylem on the inside of the ring, and secondary phloem on the outside, pushing the primary xylem and phloem apart. The vascular cambium usually consists of two types of cells: **Fusiform initials** (tall, axially oriented) and **Ray initials** (smaller and round to angular in shape)

Functions

- Fusiform initials are helpful in the formation of secondary xylem and secondary phloem components and thereby contributes in secondary growth.
- Ray initials are helpful in the formation of xylem rays and phloem rays and helps in lateral conduction of water and mineral salts through these ray initials.

Fusiform Initial Cells

These cells are highly elongated and tapering and give rise to the vertical system of a plant. They are prismatic in the centre and wedge-shaped towards the ends. In TLS (transverse longitudinal section) they appear flat and prism shaped. The derivatives of fusiform initials always divide periclinally and give rise to the axial system. Maturation of fusiform initials show characteristic growth patterns, wherein their elongation keeps on increasing, reaches a plateau and eventually dies. However in some species of *Pinus* plateau has not been reached even after 2200 years. Mean length of fusiform initial is higher in angiosperms with primitive vessel members than those with advance vessels. Storied cambia have shorter fusiform initials than non-storied cambia. Fusiform initials are the only cell types that divide along their longitudinal axis.The products formed by division of fusiform initials are: tracheids, vessel members, xylem fibres xylem parenchyma, which together forms the secondary xylem. They also gives rise to components of secondary phloem namely, phloem parenchyma, phloem fibres, sieve cells, albuminous cells (strasbuger's cells) sieve tube members along with companion cells. Number of vessel members is equal to the number of fusiform initials. Every contagious, adjoining neighboring vessel member has a common perforation plate.

Tangential Section

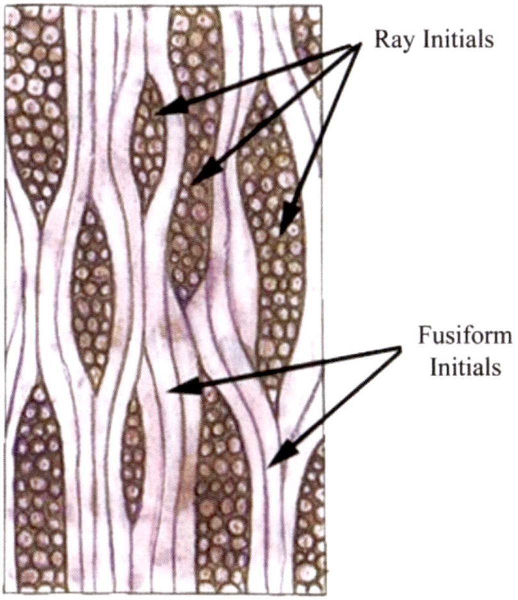

Vascular Cambium

Fig. 8.4. Ray initial and Fusiform initials.

Ray initials

Ray cells are isodiametric and upon division lead to formation of the radial system of secondary growth. They run parallel to the radii which pass through the centre of the organ.They affect the transport efficiency of water, mineral nutrients and photo-assimilates which ultimately increase the plant height. Transformation of ray initials into fusiform initials and vice versa.Fusiform and ray initial cells may interconvert into the other. Formation of ray initial from fusiform initial may be brought about by lateral partitioning, direct partitioning and septation. Reverse pattern of transformation of ray-initial to fusiform initial is although present, but uncommon.

Anticlinal and Periclinal Cell Division

All living things, including plants and living creatures are made up of one cell or multiple cells. **Cell division** is the process by which the parent cell divides into two or multiple daughter cells. Anticlinal and periclinal cell division are both different ways of cell division

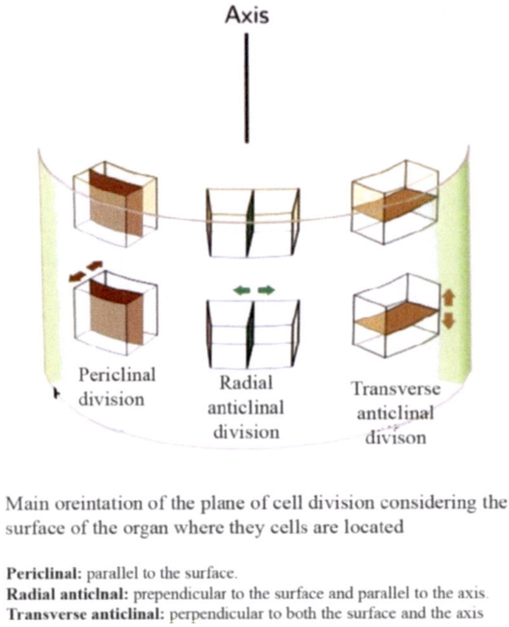

Fig. 8.5. Anticlinal and Periclinal cell division.

Periclinal cell division

Periclinal cell divisions are the ones that occur parallel to the tissue or organ surface. As a result, we get rows of cells stacked one over the other. Periclinal is parallel to the surface. Such a division results in an increase in girth of the organ thus adding length. In cylindrical organs, such as stems and roots, the term tangential may also be used in place of periclinal.

Anticlinal cell division

Anticlinal cell divisions are perpendicular to the adjacent layer of cells. So, what you get is columns of cells adjacent to one another. Anticlinal division is at right angles to the surface. The anticlinal wall of a cell is arranged perpendicular to the surface of the plant body. An anticlinal division leads to the formation of anticlinal walls between daughter cells. Such a division allows the tissue to increase its circumference, thus increasing the girth of the organ.

In simple words, anticlinal division adds more thickness and periclinal division adds length.

Physiological significance of wood formation

Wood formation is an all-inclusive term for a series of biological processes that can be arbitrarily divided into four developmental phases :

i. Awakening of the cambium from dormancy,

ii. Cell division of cambial initials and mother cells,

iii. Differentiation of cambial derivatives,

iv. Maturation.

Sapwood is physiologically important because it serves as the principal avenue/role for transformation of water and minerals, while the living cells carry on metabolic process and store food. In contrast, the dead heart wood is physiologically inactive. The function of heart wood is no longer conduction of water, but simply to give the mechanical support to the stem. The most critical change during conversion of sap wood to heartwood is the death of ray and axil parenchyma cells other important change include a decrease in the metabolic rate & enzymatic activity, starch depletion darkening of xylem. Anatomical changes such as an increase in aspiration of pits in gymnosperm and formation of tyloses in angiosperms and changes in moisture content

9

Macroscopic Features of Wood

The macroscopic characteristics of wood give the clues to the conditions under which the wood is grown, provide an indication of physical properties and serve as an aid in wood identification.

Growth rings do not always appear as distinct alternating bands of earlywood and latewood. Some hardwoods, for example, form large-diameter pores early in a growing season and much smaller pores later in the year; such woods are called ring-porous. Discontinuous rings are occasionally found in trees having one-sided crowns and in heavily defoliated, suppressed, and over mature trees. Examination of a stem cross-section often reveals a dark-colored center portion surrounded by a lighter-colored outer zone. The dark center area is known as heartwood and the lighter tissue as sapwood. Trees in which fibers are spirally arranged about the stem axis are said to have spiral grain.

Wood can often be identified by macroscopic characteristics, particularly by colour, gloss, odour, weight and structure. Wood have two types of features:

A) Macroscopic or Gross feature

B) Minute structure feature

Gross features of wood

Features can be easily seen with unaided eye may be considered as gross features.

1. Bark
2. Sap wood & heart wood
3. Pith
4. Growth rings
5. Wood rays
6. Resin or gum canals
7. Early wood and late wood

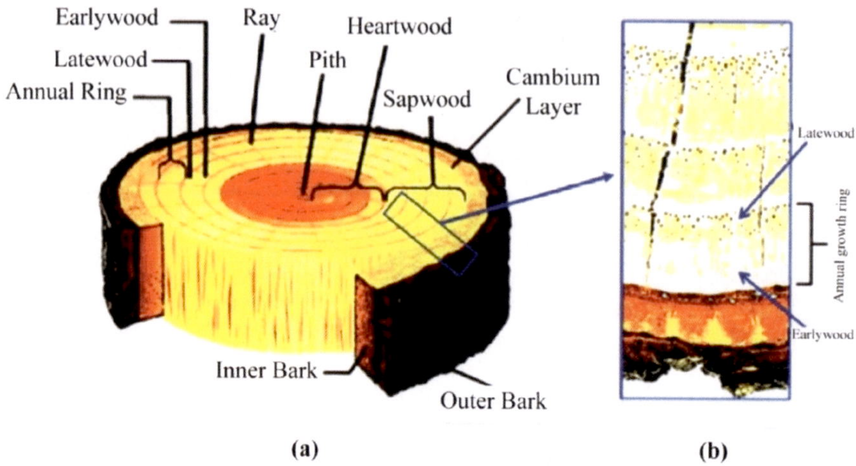

Fig. 9.1. Wood Features.

Bark

Bark varies in thickness, colour and appearance in different species. It usually show two distinct regions: an outer protective zone of crooky nature consisting of dead and dying cells and an inner living portion which is actively associated with conduction and storage of food materials required for growth. The outer bark may be smooth as in *Ficus* or it may show fissures or cracks, as in *Shorea*. Sometime the general appearance of the external bark due to fissuring or exfoliation may be distinctive as to be of some diagnostic value in the identification of trees or logs.

Sap wood & Heart wood

The woody cylinder show, as a rule , two well marked zones: an outer, lighter coloured portion known as sapwood and an inner or central dark-coloured portion known as heartwood. Colour distinction, however, is not always a true criterion. For example, in Fir (*Abies pindrow*) and Spruce (*Picea smithiana*), there is no colour distinction between the heartwood and sapwood, although, physiologically, heartwood is always present in mature wood of any tree species. Some trees like Mango (*Mangifera indica*), Semul (*Bombex ceiba*) and Fig (*Ficus* spp.) do not always show any distinct heartwood in Sal (*Shorea robusta*) show a clear cut distinct between sapwood and heartwood.

Physiologically, heartwood is dead and does not take any active part in the life of tree except to give it rigidity. All cells in heartwood are dead or skeleton cell, consisting of cell walls only without living contents. The sapwood however is composed mostly of licing cell and as its name implies, is primarily concerned with conduction of sap. The sapwood is also rich in starch and other food materials and is not durable, being easily attacked by insects and rot in the felled trees. Heartwood, on the other hand is comparatively more durable.

Comparison: Heartwood & Softwood

S. No.	Heartwood	Sapwood
1	Heartwood is produced after of a few years of secondary growth.	This is also produced due to secondary growth.
2	Most of the heartwood nears the center of the axis forms a dark coloured region is called heartwood or duramen.	There is a small outer region, however, remains light coloured is known as sapwood or alburamen.
3	Heartwood is formed due to accumulation of organic compounds, such as oils gums, resins, etc.	Sapwood is lighter coloured and it is the outer region of the secondary xylem. It is formed due to the cambial activity of the secondary xylem.
4	The heartwood is dark coloured and non-functional. The major function of this wood is mechanical support.	The cells of the sapwood are functionally active. So it helps in conduction of water.
5	Amount of heartwood increases as the tree grows older.	The amount of sapwood, however, remains almost constant.
6	The tracheary elements are plugged by tyloses.	The tracheary elements are not plugged by tyloses.
7	Tracheary elements have deposition of tannins, resins, gums, etc.	Tracheary elements do not possess any deposition in their lumen.

Pith

Pith is the soft core found near about the centre of the log. It may be round, oval, triangular or squarish in shape and is usually not more than 0.6 cm in diameter. The colour is commonly some shade of brown, though in some trees it may be pinkish, grayish or yellowish. The pith does not seem to have any specific function to perform in the mature tree. The squarish pith with concave sides is found Teak (*Tectona grandis*).

Growth rings

These are concentric ring which usually represent the wood formed during one growing season; also called, though less aptly, annual rings. These vary greatly in width, being 5 to 25 mm in fast grown timber and narrow in slow grown timber, 8 rings or more per cm. In many timbers, these are not distinct because there is little variation in their growth, throughout the growing season and a year s growth, imperceptibly merges into the next year's growth. The greater the difference between the late wood and early wood, the more prominent is the growth ring, as in temperate region.

Wood rays or medullary rays

These are group of horizontally arranged cells, running radially from bark towards the centre of the tree on radial and tangential surface often too fine to be visible. Their function is to take up the rising sap and descending food which is elaborated in the leaves, in driblets and pass it radially to the growing layer where it is required. Wood rays are organs for radial conduction and for storage of food in the trees. On the transverse section, they appear as radial lines; in tangential section short as vertical line and on the radial section as narrow horizontal ribbons. The rays appear, as small plates and produce the decorative effect of Sliver grain.

Intercellular canal (Resin canals and gum ducts)

Intercellular canals are long tubular cavities found in wood, which serve as repositories for waste products of metabolic activity like resin or gums. They are therefore, commonly referred to as resin canals or gum ducts. They may run either vertically or in the horizontal direction and are found in both porous and non-porous woods.The vertical canals can be distinguished from the vessels by the fact that they are fewer and have no walls of their own. The horizontal canals are similar to the vertical one but run in a horizontal manner through the rays. Resin cavities are intercellular spaces which arise, through pulling apart of cells and or by liquefication of cell wall substance. Pine and Spruce have both vertical and horizontal canals. The dark streak on the faces of Pine boards are due to these. In Sal (*Shorea robusta*) and Toon (*Toona ciliata*) gum ducts are found in bands of variable length at irregular intervals.

Transformation of Sapwood to Heartwood

Wood cells are produced by a layer of cells between living bark (phloem) and the wood. This layer is called the cambium and can be seen only with a microscope. Cells which form on the inside of the cambium become wood and those on the outside becomes phloem.Once wood cells are laid down by the cambium, they rapidly differentiate in size and shape until finally a thickened wall is formed, the process being completed within a milli metre or so from the outer wood layers.After the thickened wall is produced, the cells remain largely unaltered in shape for the life of the tree. The wood cells thus produced by the cambium become the sapwood.

The ray and axial parenchyma cells remain alive in the sapwood, while the more specialized wood cells, the fibres, vessels and tracheids, lose their living contents soon after differentiation and thickening of the cells are completed. Changes in the nature of the sapwood occur with the death of the parenchyma cells and with the completion of these changes heartwood is produced. These changes occur gradually through a layer of cells of variable width known as the transition zone. In this zone, increased amounts of so called tannins (polyphenolics) and coloured materials of many types are deposited in the ray and axial parenchyma. These materials diffuse into the surrounding tissue. The development of heartwood colour is due to this deposition.The development of heartwood colour is the most spectacular change between sapwood and heartwood. While a great number of timbers do not produce coloured heartwood extractives, the large majority of species show some evidence of doing so.

Heartwood colours range from white to cream as in Southern Silver Ash, Alpine Ash and European ash, to black in ebony, olive-green in lignum vitae, yellow-orange in osage orange, brown in tallo wood and red in red ironbark. Most of the colours undergo a change with age, becoming darker and more brown. Frequently the heartwood substances are unevenly distributed giving rise to bands of wood of different colour.Such banded or striped effects are evidently related to the growth conditions as they occur concentrically with the growth rings. In hard woods, as well as the production of tannins and other materials, changes occur in the vessels which become blocked on heartwood formation.This blocking of the vessels takes place by means of the pits. Pits occur abundantly in all wood cell walls and are well-defined thin regions in the wall through which solutions can readily diffuse from one cell to another.

In species with larger pits, blockage of the vessels occurs as a result of the growth of adjacent parenchyma cells through the pit forming balloon-like structures in the vessels. These growths are known as tyloses.In species with small pits, blockage of the vessels occurs due to the secretion by the ray parenchyma of tannin or gum-like materials through the pits. This deposition of materials can also occur in the species with large pits along with tyloses. Tyloses are common in the *Eucalypts* and the Merantis, whereas tannin-like materials are common in the mahoganies, true Australian Red Cedar and Queensland Maple.

In softwoods, blockage of the tracheids on heartwood formation can occur either by closing of the pits (aspiration) or by the increased deposition of resins and tannins.The pit membrane of a tracheid is a specialized structure consisting of a thickened centre (torus)-supported by a cellulose web, the web being flexible and allowing the torus to move and close (aspirate) the pit opening. Aspiration is confined to the pits of the earlywood tracheids, latewood pits remaining open.Latewood tracheids appear to be specialised structural cells not conducting cells. They have narrow lumens and thick cell walls and are formed near the end of the growing season. Earlywood tracheids have wide lumens and thin cell walls and are formed in the first part of the growing season.

Factors Affecting Transformation

The width of the sapwood depends primarily on the longevity of the parenchyma cells. It is subject to great variation between species and minor variation between individuals of the one species and even within the individual. While the sapwood width seems regular along the grain , it is apt to show fluctuations around the growth ring. Variation in sapwood width within a species has been related to many factors, some of which are:

1) **Rate of growth:** It seems fairly evident that the width of sapwood would be wider in dominant than in suppressed trees.
2) **Environment:** It also seems fairly evident that the width of sapwood would be wider in the more favourable environment; however, it has been suggested that heartwood formation is encouraged by favourable water relations in radiata pine.
3) **Position in tree:** Sapwood widths remain fairly constant with height. It can be seen therefore, due to the natural taper of a tree, that the proportion of sapwood to heartwood increases with height.
4) **Age:** Young trees have wider sapwood than old trees. It is important to note that once the tree has commenced to lay down heartwood, the volume of sapwood in proportion to heartwood becomes less as the tree becomes older. The amount and rate of heartwood formation vary greatly with species, tree age, rate of growth, environment and silvicultural practices.

10

Microscopic Features of Wood

The microscopic features reveals that wood is composed of minute units called cells. The fundamental differences between different kinds of woods are founded on the types, sizes, proportions, pits and arrangements of different cells that compose the wood. Softwoods have a simpler basic structure than do hardwoods due to the presence of only two cell types, and relatively little variation in structure within these cell types. Hardwoods have greater structural complexity because they have both a greater number of basic cell types and a far greater degree of variability within the cell types. In each case, however, there are fine details of structure that could affect the use of a wood.

The structure of a typical softwood is relatively simple. The axial or vertical system is composed mostly of axial tracheids, the radial or horizontal system and the rays, are composed mostly of ray parenchyma cells. According to estimates, 1 cubic metre (about 35 cubic feet) of Spruce wood contains 350 billion to 500 billion cells. The basic cell types are called tracheids, vessel members, fibres, and parenchyma

The structure of a typical hardwood is much more complicated than that of a softwood. The axial or vertical system is composed of fibrous elements of various kinds of vessel elements in various sizes and arrangements, and axial parenchyma cells in various patterns and abundance. Like softwoods, the radial or horizontal system are the rays, which are composed of ray parenchyma cells, but unlike softwoods, hardwood rays are much more diverse in size and shape.

The minute structure of wood can be studied under a compound microscope. When so examined; wood is seen to be cellular in structure that is made up of minute components or cells, each with a distinct wall of its own, which consists mostly of cellulose and lignin. When these cells are first formed, either at the growing points or from cambium, they are all similar in shape and size , with thin wall and living protoplasmic contents. Very soon, as a result of growth, differentiation and cell wall thickening, they become modified in size, shape and structure according to the function they have to perform. Though man in his ingenuity has found many uses for wood, in nature it has only three main functions, conduction, mechanical support and storage. Depends upon their function and structure, the various types of cells that go to make up wood are classified as below:

1. **Prosenchymatous elements**: Tracheids, vessels, fibers.
2. **Parenchymatous elements:** Parenchyma and rays, resin canals, gum canals, latex canals, infiltrants in wood.

A) Prosenchymatous elements

Vessels or Pores

Vessels are vertical series of cells with open ends place one above the other, forming a continuous tube, like the section of a drain pipe, running in the direction of the long axis of the tree. Their function is to conduct sap (water & mineral nutrients) from the soil and roots to the crown for which they structurally well adapted.On the longitudinal surfaces or board faces of timber, the vessels frequently show up as long fine scratches or grooves, especially in coarse-textured wood (Kokko and Mango). When cut across, they appear on the end surface of wood as small, circular or somewhat oval opening or holes, usually visible to the eye or hand lens and are therefore also known as pores. The occurrence of pores or vessels is a constant feature of all broad-leaved trees, and it is for this reason that the wood of broad-leaved species is described as porous. On the other hand, in conifers (softwood), vessels or pores are entirely absent and the wood is accordingly classified as non-porous. The arrangement and size of vessels vary in different species which fact is of diagnostic value in the identification of hardwoods.

Fibres

Fibres are vertically aligned, narrow, elongated and thick-walled cells with pointed tapering ends, which make up the bulk of the wood by weight in most hardwoods. Their main function is to give mechanical support to the tree, and like the vessel, they are absent in non-porous or coniferous woods. The fibres are individually indistinct even under a lens, but collectively they form, as it were, the darker coloured ground mass in which the vessels and other lighter-coloured wood elements are arranged, giving rise to various pattern.

Tracheids

These are longitudinal elements, hollow, needle-shaped, resembling fibres and characterized by the presence of large bordered pits on their walls. Tracheid cells are minute and narrow thick walled with closed tapering ends, closely packed together so that a cross section through them resembles a honey-comb. These are the main longitudinal elements of conifers or non-porous wood.

Their main function is to conduct sap and to give rigidity to the tree.

B. Parenchymatous elements

I. Parenchyma or soft tissues

These are short, rectangular or brick-shaped cells, which comparatively thin walls and simple pits, whose main function is storage and conduction of food material. They are oriented in the same direction as fibres and vessels, with their long axis parallel to the grain. Therefore, sometimes they are also referred to as longitudinal parenchyma whereas the parenchyma cells of the rays are arranged horizontally. They are found in both porous and non-porous woods. Individually the longitudinal parenchyma cells are too small to be seen under a hand lens, but collectively they form various patterns which are often very distinctive and of considerable importance in the classification and identification of timber.

Parenchyma described under two classes

1. Apotracheal

These are independent of the distribution of the pore

A) **Terminal or initial**: In some hardwoods, the parenchyma cells are arranged in a continuous line or narrow band, which may be formed either at the beginning or at the end of the growth season. Such bands of soft tissue, which are usually lighter coloured than the background, are classified as initial or terminal, depending upon as they are formed.

B) **Diffuse:**When soft tissues occur as isolated cells or small groups of 2-3 cells scattered throughout the ground mass of the wood.

C) **Reticulate or net-like:** In some wood, the parenchyma cells, instead of being scattered, may form fine, more or less evenly spaced, tangential lines, which together with the rays give rise to a characteristic pattern somewhat resembling the meshes of a net. Such a distribution is described as reticulate or net-like. The parenchyma lines may be broken or continuous, closely to somewhat widely-spaced and of varying of thickness, giving rise to further variation in reticulate pattern.

2. Paratracheal

In these soft tissues are closely associated with the vessels

A) **Vasicentric:** The soft tissues form a narrow but complete sheath, more or less of uniform thickness all around, surrounding the pores. The parenchyma sheath is distinctly visible under lens, appearing as a light-coloured border or halo round the pores as in Babul (*Acacia nilotica*).

B) **Aliform or eyelet type:** The parenchyma surrounding the pores, instead of being uniformly thick and narrow, may extend sideways as wing like lateral extensions. Such a distribution of parenchyma round the pores is described as Alifom or eyelet eg. *Acrocarpus fraxinifolius* and *Mangifera indica*. A further development of the Alifom type, in which the wing-like extensions of adjacent pores become confluent of connected together, is known as Aliform confluent eg. *Dalbergia sissoo* & *Ougeinia oojeinensis*.

C) **Banded:** When soft tissues are arranged in continuous tangential bands, alternating with fibre layers throughout the growth ring. The bands may be relatively broad or narrow, straight or wavy, and confluent or independent of the pores eg. *Butea monosperma* & *Ficus* spp.

II Resin canals

Resin canals or **resin ducts** are elongated, tube-shaped intercellular spaces surrounded by epithelial cells which secrete resin into the canal. These canals are orientated longitudinally and radially in between fusiform rays. They are usually found in late wood: denser wood grown later in the season. Resin is antiseptic and aromatic and prevents the development of fungi and deters insects

Types

a) **Normal resin canals:** Normal resin canals exist naturally in the wood of the genera *Picea. Larix, Pinus,Pseudotsuga* and *Shorea.*

b) **Traumatic resin canals:** Traumatic resin canals may be formed in wounded trees that don't have normal resin canals. Wounding occurs from either fire, freezing or mechanical damage. These canals are irregularly shaped compared to normal resin canals

Gum Ducts (Canals)

Gum ducts are produced as a result of wounding in many hardwoods and such canals are said to be traumatic; they may be distinguished from the normal type because they are invariably in tangential series, are irregular in outline, contain no epithelial cells and usually have dark-coloured, more-or-less viscus, gum like deposits. Further, traumatic canals are usually larger than the vessels and typically wider tangentially. In some species, such as African Walnut, traumatic canals are sufficiently frequent in occurrence to be regarded almost as a characteristic feature of the timber.

Latex canals or tubes

Special cells, containing latex and called latex tubes, occur in the ray tissue of certain timbers. They are usually invisible to the naked eye, but where they can be detected a helpful feature in identification. In a few timbers (Jelutong and Mujwa) specialised parenchymatous tissue, containing numerous latex tubes, develops from leaf-traces and continues outwards during the subsequent growth of the bole, such latex traces, as seen on the tangential surfaces, are upto 20 mm high and lens-shaped section. As the leaf –traces occur in whorls, the latex traces are found in tangential series up the tree at intervals of 0.6-0.8 metres, disfiguring long lengths of timber, and rendering it unsuitable for many purposes. Long splits can develop from the latex traces during seasoning.

11

Physical Properties of Wood

Physical properties of wood

The versatility of wood is demonstrated by a wide variety of products. This variety is a result of a spectrum of desirable physical characteristics or properties among the many species of wood. In many cases, more than one property of wood is important to the end product. For example, to select a wood species for a product, the value of appearance-type properties, such as texture, grain pattern and color, may be evaluated against the influence of characteristics such as machinability, dimensional stability and decay resistance.

Wood exchanges moisture with air; the amount and direction of the exchange (gain or loss) depend on the relative humidity and temperature of the air and the current amount of water in the wood. This moisture relationship has an important influence on wood properties and performance.

1. Colour
2. Hardness
3. Weight
4. Texture
5. Grain
6. Lustre
7. Odour
8. Figure

Colour

The unique colour of a wood species is determined by the chemical extractives embedded in the cell walls. But the initial colour of the raw, freshly cut wood doesn't remain unchanged. This colour darkens somewhat when you apply a finish, even if that finish appears clear and colourless. Most finishes also change the hue, making it more amber. Craftsmen describe this as "warming up" the wood colour. In the chart of "Physical Properties of Wood," you can see how common finishes affect colour. Unfortunately, what you can't see is that the wood also changes colour with age. As the surface of the wood is exposed to air, it slowly oxidizes. Some woods are photosensitive — exposure to ultraviolet light alters the extractives. Both reactions change the wood colour at the surface.

Hardness

It signifies the resistance offered by a wood specimen to penetration by another body i.e. resistance to indentation and not cutting which depends on nature of grain, presence of silica and other substances.

Factors responsible for hardness in wood:

i) **Anatomical structure:** The degree of cohesion of fibres determines the hardness. Heartwood is harder than sapwood.

ii) **Specific gravity:** Generally, hardness increases the specific gravity i.e. slow grown coniferous timber is harder than fast grown wood of the same species.

iii) **Moisture content:** Dry wood is harder than green wood.

iv) **Presence of resinous material:** Increase the hardness.

v) **Soundness:** The decay wood softens by disintegrating tissues.

Sr.No.	Category	Example
1.	Extremely hard	*Shorea robusta, Acacia nilotica, Acacia catechu* and *Dalbergia latifolia*
2.	Very hard	*Dalbergia sissoo, Morus alba, Terminalia alata* and *Syzygium cuminii*
3.	Hard	*Albizia lebbeck, Tectona grandis, Adina cordifolia* and *Churasia tabularis*
4.	Soft	*Lannea coromandelica, Cedrus deodara* and *Abies pindrow*
5.	Moderately soft	*Mangifera indica*
6.	Very soft	*Boswellia serrata, Pinus roxburghii, Picea smithiana* and *Pinus wallichiana*
7.	Extremely soft	*Bombax ceiba* and *Cryptomeria japonica*

Weight (Specific gravity)

Weight of wood varies with species and even in different parts of the same tree. The density of any material is defined as the weight of an unit volume of the material and is expressed as kg/cum. The specific gravity, is the ratio of the weight of material to the weight of an equal volume of water at 4° C. Specific gravity in itself is only of secondary importance; but it is closely connected with other properties such as strength, durability, soundness, combustibility and seasoning power.

Specific gravity of wood depends on:

1. **Anatomical structure:** The thicker the cell walls, the greater the specific gravity. Slow grown wood is heavier than fast grown wood.
2. **Soil and locality:** The weight depends on conditions of soil fertility, moisture, heat and light. Excess of moisture renders wood porous, and spongy hence reduce specific gravity. Insufficient light produces similar results. Soil fertility also increase the weight.
3. **Age of the tree:** The formation of heavy heartwood during middle age may render the average specific gravity of wood of older tress higher than of young trees.

Texture

Texture pertains to the relative size of cells and their proportion in unit volume. Texture may be fine, coarse, even or uneven and with intermediate shade. Fine textured wood take polish easily, even textured wood are less decorative than uneven textured one. Texture depends on the size of the longitudinal cells. In hardwoods, this refers to the size of the **fibers** and the **vessel elements**; in softwoods, it's the **tracheids**. Fine-textured woods have small cells, while coarse woods have larger ones.

Grain

Grain pertains to alignment of cells or direction of fibres with regards to vertical axis of the tree. When the vessel elements in hardwoods are sliced open, they create depressions called **pores**. Woods with large pores that are easily visible to the naked eye are said to have an **open grain**. Those with smaller pores, to small to see clearly, have a **closed grain**. Softwoods do not have vessel elements, these terms don't apply for practical purposes – such as preparing and finishing the surface – softwood grain is closed.

Grain Pattern

The normal orientation of the longitudinal cells determines a species' characteristic grain pattern. There are four categories.

Straight grain: The longitudinal cells grow fairly straight and parallel to the axis of the trunk.

Wavy grain: The cells undulate in short, even waves. This sometimes produces curly figure.

Irregular grain: The cells undulate around knots or in no discernible pattern.

Interlocked grain: The cells spiral around the trunk, reversing direction every few growth rings. This produces ribbon figure.

Diagonal: The direction of fibres is not parallel with the longitudinal axis of the board or plank.

Spiral: Fibres follow a spiral course.

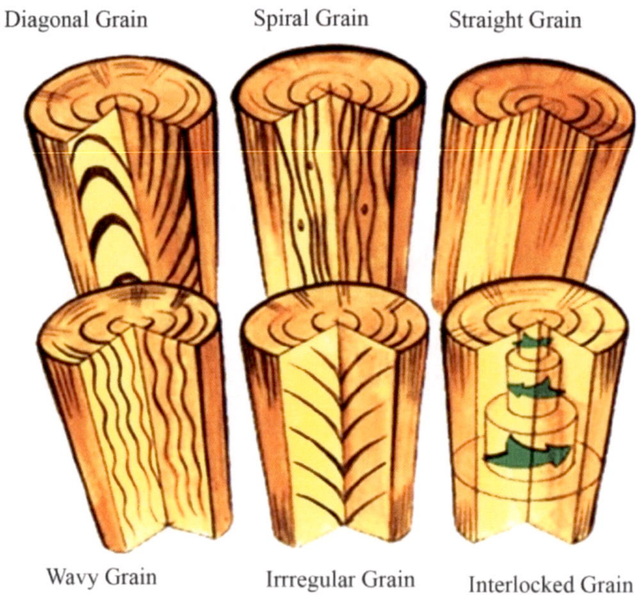

Fig. 11.1. Types of Grain

Lustre

Lustre or sheen depends on the ability of the cell walls to reflect light. This property is not present in all timbers. Lustre is conspicuous on longitudinal and radial section spp., where maximum width of ray is exposed eg. *Picea smithiana, Quercus* spp. *Albizia lebbek,* etc.

Odour

Tannins fatty acids, act as bearers of scent in wood. Many timbers have characteristics odour, particularly when freshly cut. Odour is more pronounced in heartwood than sapwood. Resinous odour in Pines, spicy aroma of Sandalwood (*Santalum album*), unmistakable odour of deodar (*Cedrus deodara*), Camphor (*Cinnamomum camphora*) and Teak (*Tectona grandis*).

Figure

Figure refers to the distinctive pattern produced on longitudinal surfaces of timber as a result of arrangement of different tissues and direction of the grain. Figure may be attractive due to differences in the reflective property of different surfaces.

C) According to condition

1) **Live knot**: a knot free from decay and other defects, firmly intergrown with surrounding wood (intergrown knot, tight knot and sound knot)
2) **Decayed knot:** a knot soften than the surrounding wood and containing decay (unsound knot, punk knot and rotten knot)
3) **Loose knot:** a knot which is not held firmly in place by growth or position.
4) **Dead knot:** a knot in which the wood elements are not comptletely intergrown wit those of the surrounding wood (encased knot and black knot)
5) **Pith:** a sound knot with a pith hole not over ¼ inch (0.6cm) in diameter
6) **Enclosed knot:** a knot that does not appear on the surface of timber (a hole left as a result of the removal of a knot is called a knot hole)

Pith

The pith is an essential part of the living tree, but from a wood utilization point of view it is defect. Its presence in wood products may reduce both strength and durability. This is primarily due to the different cellular structure of the pith in comparison to the surrounding wood. In addition the wood immediately around the pith often contains small checks and knots of various sizes, and may also develop shakes. In sawing logs to lumber, the pith is left enclosed within (boxed-heart) timbers or it may be sawn through and removed.

Pith flecks

Certain timber may often show irregular batches of soft tissue, which seem to have no relation whatsoever to the general arrangement of parenchyma. These abnormal patches of parenchyma are formed as a result of injury to the cambium due to insect attack and known as pith flecks. On the board face, they show up clearly against the natural colour of the wood, but do not constitute a serious defect unless of very large size and numerous. Some species may be more liable to form pith flecks than others.

Three Dimensional Features of Wood; Transverse, Tangential and Radial Surfaces.

The structure of wood is three-dimensional due to the arrangement of the layers of growth in the tree, as well as the vertical or horizontal orientation of the individual cells.

Planes of Section

Although wood can be cut in any direction for examination, the organization and interrelationship between the axial and radial systems give rise to three main perspectives from which they can be viewed to glean the most information. These three perspectives are the transverse plane of section (the cross section), the radial plane of section, and the tangential plane of section. Radial and tangential sections are referred to as longitudinal sections because they extend parallel to the axial system (along the grain).

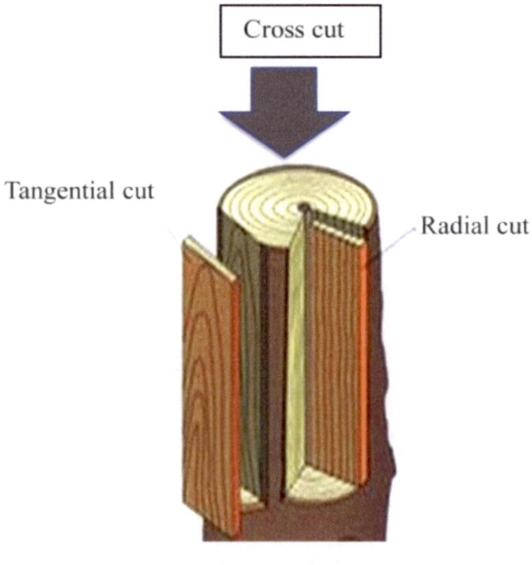

Fig. 11.2. Planes of section of wood

Transverse Plane of Section

The transverse plane of section is the face that is exposed when a tree is cut down. Looking down at the stump one sees the transverse section cutting a board across the grain exposes the transverse section. The transverse plane of section provides information about features that vary both in the pith to bark direction (called the radial direction) and also those that vary in the circumferential direction (called the tangential direction).It does not provide information about variations up and down the trunk.

Radial plane of section

The radial plane of section runs in a pith-to-bark direction and it is parallel to the axial system, so it provides information about longitudinal changes in the stem and from pith to bark along the radial system. To describe it geometrically, it is parallel to the radius of a cylinder, and extending up and down the length of the cylinder. In a practical sense, it is the face or plane that is exposed when a log is split exactly from pith to bark. It does not provide any information about features that vary in a tangential direction.

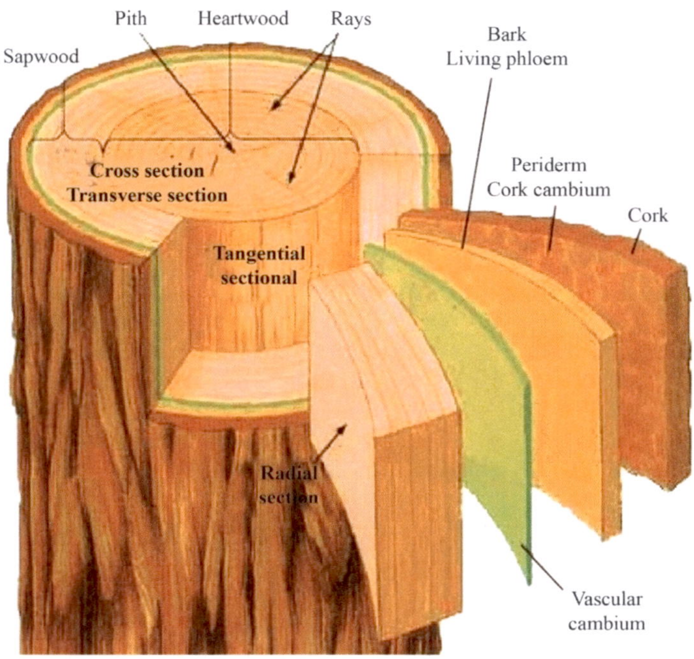

Fig. 11.3: Wood Section

Tangential Plane

The tangential plane is at a right angle to the radial plane. Geometrically, it is parallel to any tangent line that would touch the cylinder, and it extends along the length of the cylinder.One way in which the tangential plane would be exposed is if the bark were peeled from a log; the exposed face is the tangential plane. The tangential plane of section does not provide any information about features that vary in the radial direction, but it does provide information about the tangential dimensions of features. All three planes of section are important to the proper observation of wood, and only by looking at each can a holistic and accurate understanding of the three-dimensional structure of wood be gleaned. The three planes of section are determined by the structure of wood and the way in which the cells in wood are arrayed. The topology of wood and the distribution of the cells are accomplished by a specific part of the tree stem.

1) Cross Section

When a section is cut at right angle to the long axis then it is a cross / transverse section.

2) Radial Section

When a section is cut parallel to the long axis and it passes through the radius of the axis then it is radial section or radial longitudinal section .

3)Tangential Section

A tangential section is a longitudinal section of a stem where the section is perpendicular to the tangent or radius of the stem.

12

Chemical Composition of Wood Cell Walls

Cell walls mainly consist of **cellulose**, **hemicellulose**, and **lignin** in a 4:3:3 ratio. This ratio differs from sources such as hardwood, softwood, and herbs. Besides these three components, **natural lignocellulosic materials** contain a small amount of **pectin**, nitrogenous **compounds**, and the secret ash.

1. Cellulose
2. Hemicellulose
3. Lignin

Cellulose

Cellulose is a glucan polymer consisting of linear chains of 1,4-ß -bonded anhydroglucose units. Cellulose is the structural component of the primary cell wall of the green plants.About 33% of all plant matter is cellulose (the cellulose content of cotton fiber is 90%, that of wood is 40-50% and that of dried hemp is approximately 45%. Cellulose is a straight chain polymer.The number of sugar units in one molecular chain is referred to as the degree of polymerization (DP).Cellulose is insoluble in most solvents including strong alkali. It is difficult to isolate from wood in pure form because it is intimately associated with the lignin and hemicelluloses.

Hemcelluloses

Hemicelluloses are mixtures of polysaccharides synthesized in wood almost entirely from glucose, mannose, galactose, xylose, arabinose, 4-O methylglucuronic acid, and galacturonic acid residues.Some hardwoods contain trace amounts of rhamnose. Generally, hemicelluloses are of much lower molecular weight than cellulose and some are branched. They are intimately associated with cellulose and appear to contribute as a structural component in the plant. Some hemicelluloses are present in abnormally large amounts when the plant is under stress; e.g., compression wood has a higher than normal galactose content as well as a higher lignin content. Hemicelluloses are soluble in alkali and easily hydrolyzed by acids

Lignin

Lignin is a phenolic substance consisting of an irregular array of variously bonded hydroxy- and methoxy-substituted phenylpropane units. The precursors of lignin biosynthesis are: **p-coumaryl alcohol, Coniferyl alcohol**, and **Sinapyl alcohol.**

p-coumaryl alcohol is a minor precursor of softwood and hardwood lignins; **coniferyl alcohol** is the predominant precursor of softwood lignin; and **coniferyl alcohol** and **sinapyl alcohol** are both precursors of hardwood lignin. These alcohols are linked in lignin by carbon–carbon bonds.

Extractives in wood

Wood extractives, or wood extracts, can be regards as non-structural wood constituents and usually represent a minor fraction in wood or tiny molecules that are extracted from wood using solvents or other extraction methods. The extractives comprise both inorganic and organic components. The extractives are the waxes, fatty acids, resin acids, and terpenes of a tree. They are classified as phenolic, aliphatic, alicyclic, terpenoids, terpenes compounds, or other lesser compounds. The overall composition of the extracts varies dramatically from tree species to tree species and various parts of the same tree, e.g. stem, branches, roots, bark and needles, differ markedly with respect to both their amount and composition of extractives. Generally, content of extractives is higher in bark, leaves and roots, than that in stem wood. The growth environment that the tree grows within also has an effect on the composition and levels of wood extractives.Wood extracts are considered to have a low molecular weight, and do not have an effect on the growth rate of the plant. Therefore, they are not vital to plant growth.

Role of extractives

Extractives can effect the density and strength of the wood species. The more extractives that are embedded in the fibers, the denser and stronger the wood. Many extractives are minerals. These act as fine abrasives, dulling your cutting tools as you work.

Some are resinous and waxy substances. These build up as pitch on the cutting surfaces of power tools and interfere with the cutting action. They can also make a wood difficult to glue or finish. Many are antibiotics, killing molds and bacteria that would cause disease and rot. Woods high in these chemicals are resistant to decay. A few of these antibiotic chemicals are toxic to humans as well as microbes and have been known to cause or aggravate allergic reactions, respiratory ailments, and other health problem

13

Anatomical Features of Common Indian Timbers

Ring Porous

- having vessels more numerous and usually larger in cross section in the springwood with a resulting more or less distinct line between the springwood and the last season's **wood**— compare **diffuse-porous**.
- **In some species** (e.g. Oak (**Red Oak**, **White Oak) Teak** and ash **elm**, **hickory**), the largest pores are in the earlywood while those in the latewood are more evenly distributed and uniform in size.
- These woods typically have distinct figures and patterns, and the uneven uptake of stain (the large pores soak up more colour) make the figure more pronounced.
- These are also known as open-grain woods

Semi-ring porous or Semi-diffuse porous

- In some species (e.g. Butternut, **Black Cherry**, **Black Walnut**, **Pecan**, **Tanoak**), pores are large in the earlywood and smaller toward the latewood, but without the distinct zoning seen in ring-porous woods.
- Also, some species that are usually ring-porous (e.g. Cottonwood) occasionally tend toward semi-ring porous.

Diffuse Porous

- In some species (e.g. Maple, Cherry, Yellow poplar **Basswood**, **Red Alder**, **Sugar Maple**, **Sycamore**, **Yellow Birch**, **Yellow Poplar**.) the pores are distributed fairly evenly across the earlywood and latewood.
- Most domestic diffuse-porous woods have relatively small-diameter pores, but some tropical woods of this type (e.g. Mahogany) have rather large pores.
- These woods usually have even uptake of stain (there seems to be no

scientifically proven explanation of the cause of blotching). These are also known as closed-grain woods.

Non-porous

- Softwoods don't have vessel cells (water is conducted in the living tree in tracheid cells).
- Different softwoods have different growth-ring characteristics.
- In White Pine, the rings are non-distinct, and stain uptake is fairly even, as in diffuse porous woods.
- In Yellow Pine, where the rings are clearly visible, stain uptake in earlywood is more pronounced than in latewood, as in ring-porous woods.

Juvenile wood and mature wood

Juvenile wood

- Wood formed near the pith, characterized by progressive increases in dimensions.
- Changes in the cell characteristics, and the pattern of cell arrangement; also called core wood.
- Secondary xylem at the center of a tree formed throughout the life of the tree.
- Pith Plus the wood produced for the first 2 to 20 year
- The juvenile wood zone is conically shaped area that narrows from a broad base toward the top of the tree.

Mature Wood

- Wood which is characterized by relatively constant cell size, well-developed structural
- Patterns, and stable physical behavior, also called adult wood.
- Occurs as the cambium becomes farther from and less influenced by the apical meristems.

Anatomy of Some Indian Timber Species

1. ***Acacia auriculiformis***

Trade Name: *Acacia*

Vernacular Names: Akasia (Indonesia); Australian Babul, Australian Wattle, Acacia, Kasia (India); Darwin Black Wattle and Tan Wattle (Australia)

Botanical Name: *Acacia auriculiformis* A. Cunn.ex Benth.

Family Name Leguminosae (Fabaceae)

Origin (Distribution): An exotic. Native to Papua New Guinea (PNG), Australia and Solomon Islands; introduced into many tropical countries as a fast growing species for pulp wood.

Tree Description: A medium-sized tree reaching about 20m in height and 90cm in diameter. The species has become naturalized in many parts of India including Kerala.

General features

i. *Acacia auriculiformis* is moderately heavy wood.
ii. Heartwood is dark brownish.
iii. Wood is moderately heavy.
iv. Parenchyma paratracheal and vasicentric.
v. Rays fine, numerous and closely spaced.
vi. Vessels solitary and in radial multiples of two to three.
vii. Colour : Heartwood is light brown to dark red; clearly demarcated from the yellowish white sapwood.
viii. Weight moderately heavy (Air-dry specific gravity 0.60-0.75 with average values of 0.72).
ix. Grain straight or wavy texture fine.

Gross features

a) Growth rings fairly distinct. Diffuse porous wood.
b) Vessels solitary and in radial multiples of two to three, large to medium sized and moderately numerous (15-24/mm^2).
c) Soft tissue forms sheaths around vessels.

d) Rays fine, numerous and closely spaced; one to three cell wide, homogeneous, wholly made up of procumbent cells.

e) Parenchyma paratracheal and vasicentric.

Uses: For furniture making and construction purposes. Mainly used for pulp wood production. Suitable for door and window shutters, light construction, furniture, flooring, industrial and domestic woodware, tool handles, turnery articles, carom coins, agricultural implements, charcoal, etc.

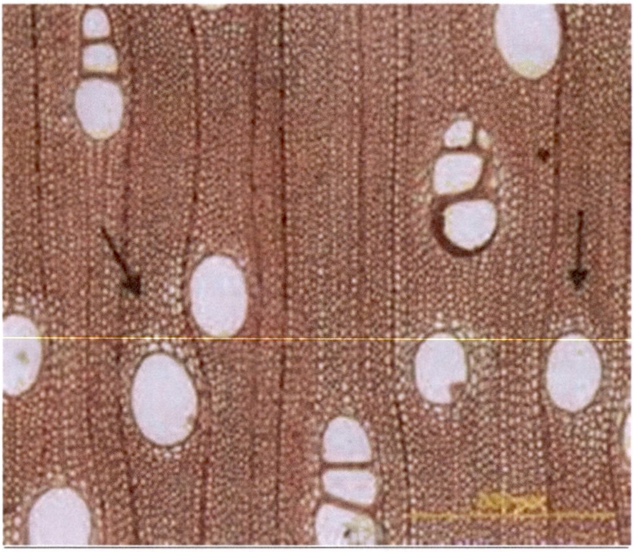

Fig. 13.1: Cross section of ***Acacia auriculiformis***

2. *Artocarpus hirsutus* Lamk.

Trade Name: Ayani

Vernacular Names: Anjili, Ayani, Ansfeni (Mal.), Hebbalesu, Hessua, Hessuain (Kan.), Dinipilla (Tam.) and Pejata (Tel.)

Botanical Name: *Artocarpus hirsutus* Lamk.

Family Name: Moraceae

Origin (Distribution): West coast tropical evergreen, West coast semi-evergreen and Southern secondary moist mixed deciduous forests; commonly seen in Western Ghats in Karnataka and Kerala and in South Maharashtra

Tree Description: A lofty evergreen tree, large to very large, 25-45m in height with a clear bole of 10-20m and 130cm in diameter. Bark dark brown and smooth.

Artocarpus hirsutus Lamk (Ayani) is a moderately hard and moderately heavy wood.

- Heartwood is golden yellow to yellowish brown.
- Parenchyma: paratracheal- vasicentric to aliform.
- Rays moderately broad to fine, fairly wide spaced.
- Vessels very large to large, few, solitary or in radial multiples of two or three, often filled with tyloses or white chalky deposits.
- Resembles with Jack wood in many respects, but differs in its yellowish-brown colour and lustrous appearance.

Colour of Heartwood is golden yellow to yellowish brown, lustrous when first exposed; darkens on exposure to dark yellowish brown with darker streaks.

Sapwood is greyish or yellowish white.

Weight moderately hard and moderately heavy (595 kg/m^3 at 12% m.c.)

Grain straight to interlocked.

Texture medium to coarse

Gross featues: Growth rings indistinct to distinct, when distinct demarcated by slightly denser and darker coloured latewood fibrous tissues.

A diffuse porous wood.

Vessels very large to large few solitary or in radial multiples of two or three, oval to round in outlines, often filled with tyloses or white chalky deposits.

Parenchyma paratracheal - vasicentric to medium to aliform, forming light coloured haloes or eye-lets round the vessels often extending sideways joining similar extensions.

Rays moderately broad to fine, fairly wide spaced and uniformly distributed, forming distinct or inconspicuous flecks on the radial surface.

Uses: The wood is used practically for all purposes for which teak is used. It is also used for furniture and cabinet work, boat and ship building, vehicle bodies, beams and rafters, windows, door frames and ceiling boards, furniture and cabinets, turnery, piles, flush door shutters, class I plywood and veneers, marine plywoods, blockboard, tool handles, beater heads, fence posts, textile mill accessories, cooperage, hurdles for sports, mathematical, engineering and drawing instruments, brushware, carts and carriages. It is also used for masts, rafts and life boats, railway sleepers, packing cases, ladder and trestles, mallet heads, tent accessories, artificial limbs and turnery articles, general decorative marine and concrete shuttering plywood, blockboards and flush doors.

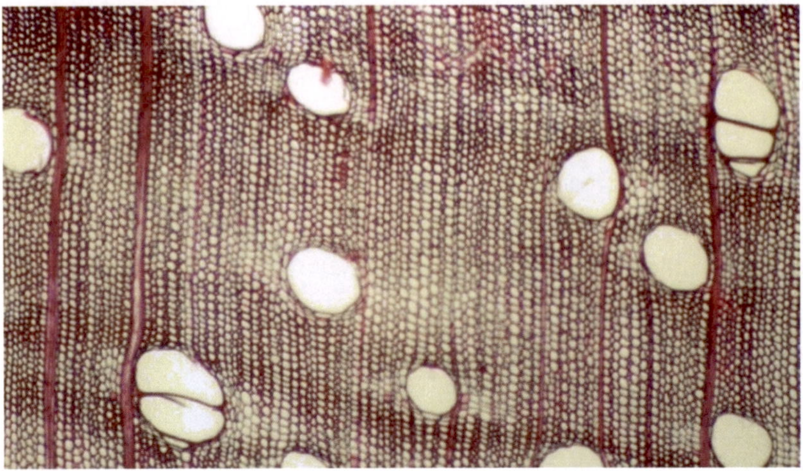

13.2: Cross section of *Artocarpus hirsutus* Lamk.

3. *Acacia nilotica*

Trade Name Babul

Vernacular Names Babul (India), Gabdi (Cameroon), Mgunga (East Africa), Babla (Bangali), Bamura (Gond.), Baval, Bawal (Guj.), Babul, Babur (Hind.), Fati, Gobli, Jali, Karrijali, Meshwal (Kan.), Karuvelam (Mal.), Kikar (Punj.), Karuvai, Karuvelam (Tam.), Nella tuma (Tel.)

Botanical Name *Acacia nilotica* (Linn.) Wild. ex Del. Syn. *Acacia arabica* Auct. non (Lamk.) Willd. **Family Name** Leguminosae (Fabaceae)

Origin (Distribution) India (Gujarat, Rajasthan, Maharashtra, Madhya Pradesh) and Africa. Grows naturally in the deciduous forests of Peninsular India. In Kerala, occasionally grown in dry areas.

Tree Small sized evergreen tree reaches a height of 10 m and is about 30 cm in diameter with short thick cylindrical trunk and a spreading crown. Bark is dark brown or black, rough, with deep narrow longitudinal fissures running spirally.

General features:

The heart wood of *Acacia nilotica* (Babul) is pinkish brown to reddish brown and is lustrous.

- Pinkish brown to reddish brown and lustrous heart wood.
- Wood is hard and heavy

- Growth rings are indistinct
- Prominent paratracheal (vasicentric) parenchyma
- Vessels are small and few and filled with dark brown gummy deposits.

Colour Heartwood pinkish brown to reddish brown, lustrous; Sapwood wide, white to pale yellow. Heart wood is sharply demarcated from the sapwood.

Weight Hard and heavy (800 kg/m^3 at 12% m. c.)

Grain Straight to interlocked

Texture Medium to coarse

Gross features

Wood is diffuse porous.

Growth rings are indistinct.

Vessels medium to small, few to moderately few, mostly solitary or in radial multiples of two, three or more, occasionally in clusters; filled with dark brown gummy deposits.

Soft tissues are visible to the eye forming thick sheaths or 'halos' round the pores. Parenchyma paratracheal- vasicentric, fine lines delimiting growth rings.

Rays moderately broad to fine, rather widely and irregularly spaced.

Uses Used in constructional work for posts, beams, rafters, bridges, door and window shutters and frames. Also used for agricultural implements, tool handles, tent accessories cart building, parts of hulls of boats, legs of bedsteads, sports goods and charcoal. The tree yields pale yellow to brown gum used in confectionery, medicines and calico printing. A heavy, somewhat twisted- grained and coarse textured reddish brown non-ornamental wood; one of the best of Indian *Acacia* species.

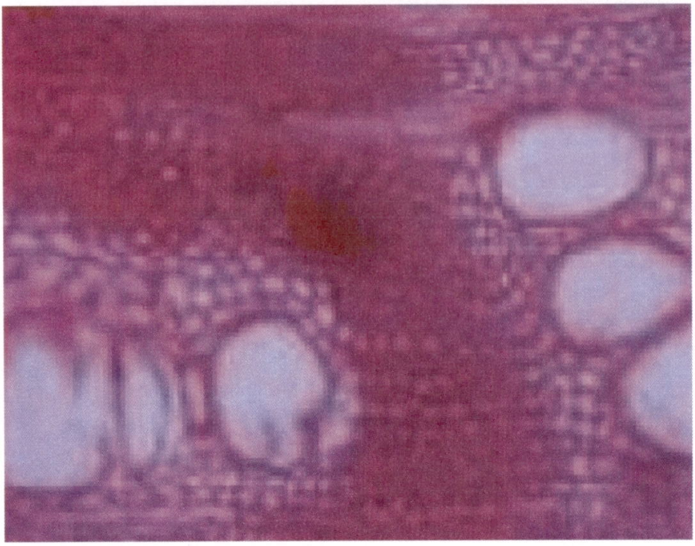

Fig. 13.3: Cross Section of *Acacia nilotica*

4. *Ficus bengalensis*

Trade Name Banyan

Vernacular Names Peepal, Peral, Banyan (India), Bor (Asm.), Vad, Vor (Guj.), Bar, Bargad (Hind.), Alada (Kan.), Vata (San.), Ala, Alai (Tam.), Marri (Tel.), Aal, Peral (Mal.)

Botanical Name *Ficus bengalensis* L.

Family Name Moraceae

Origin (Distribution) Asia-through out the forest tracts of India, both in sub-Himalayan region and in deciduous forests of Deccan and other parts of South India.

Tree Description: Large tree with spreading branches attaining a height of 30 m.

Bark greyish white, smooth.

General features

Ficus bengalensis L. (Banyan) is light to moderately heavy wood.

- Creamy white to greyish white when first exposed turning grey or pale brownish grey with age, discolours rapidly by sap stain.

- Wood is light to moderately heavy
- Parenchyma abundant, visible to the eye, forming more or less concentric narrow or broad bands alternating with fibrous tissue: 4-20 per 5mm.
- Rays moderately broad to broad, distinct to the eye, widely spaced, forming inconspicuous to conspicuous flecks on radial surface.

Colour Creamy white to greyish white when first exposed turning grey or pale brownish grey with age, discolours rapidly by sap stain; heartwood and sapwood not distinct.

- Weight light to moderately heavy.
- Air- dry specific gravity approx. 0.61
- Grain shallowly interlocked.
- Texture coarse and uneven.
- **Gross features**
- Diffuse porous wood.
- Growth rings inconspicuous.
- Vessels large to small distinct under the hand lens, solitary or in radial multiples of 2-4, few to moderately numerous (5-15 per mm^2), more or less evenly distributed, open or often plugged with tyloses, vessel lines distinct.
- Parenchyma abundant, visible to the eye, forming more or less concentric narrow or broad bands alternating with fibrous tissue: 4-20 per 5mm.
- Rays moderately broad to broad, distinct to the eye, widely spaced, forming inconspicuous to conspicuous flecks on radial surface.

Uses: Third class wood used for making tea boxes, toys and for light packing cases. The wood of aerial roots can be used for tent poles, cart yokes and carrying shafts.

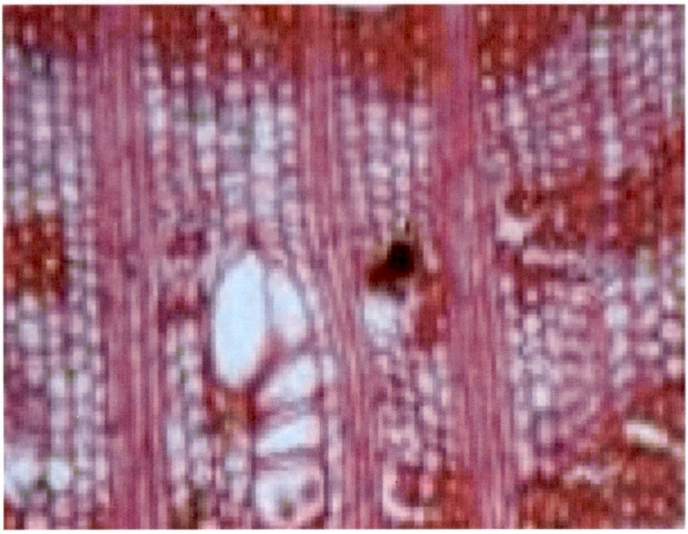

Fig. 13.4: Cross Section of *Ficus bengalensis*

5. *Lagerstroemia microcarpa*

Trade Name: Benteak

Vernacular Names: Arale, Bendeku, Brntik, Nandi (Kan.), Ventek (Mal.), Nana (Mar.), Vivella, Venteak, Venteku (Tam.), Chennangi, Ventaku (Tel.)

Botanical Name: *Lagerstroemia microcarpa* Wight. Syn., *Lagerstroemia lanceolata* Wall. ex Clarke and *Lagerstroemia thomsonii* Koehne

Family Name: Lythraceae

Origin (Distribution): West coast semi evergreen, moist teak bearing and Southern moist mixed deciduous forests. It is common throughout the Western Ghats from Bombay southwards through Karnataka extending to Travancore and the Nilgiris.

Tree Description: Large about 20-30 m in height and up to 110 cm in diameter, 2.4-3 m in girth with a clear bole of 12 to 15 m.Bark white or yellowish- grey, smooth, exfoliating in large papery flakes.

General features

- *Lagerstroemia microcarpa* Wight. (Benteak) is a moderately heavy wood.
- Heartwood reddish brown to walnut brown.
- Wood is moderately heavy.

- Parenchyma paratracheal- predominantly aliform-confluent.
- Rays fine, numerous and closely spaced.
- Vessels large in early wood, medium to small in late wood, moderately numerous (4-6/mm^2), solitary or in radial multiples of 2 or 3, usually filled with abundant tyloses.
- Colour: Sapwood grey or pink, heartwood reddish brown to walnut brown, darkening on exposure.
- Weight moderately hard to moderately heavy, slightly to fairly lustrous, 640 kg/m^3 at 12% m.c.
- Air dry specific gravity 0.59 to 0.76
- Grain straight to somewhat interlocked.
- Texture medium

Gross features

- Wood is semi-ring-porous to ring porous.
- Growth rings distinct, demarcated by large early wood vessels and fine to fairly broad parenchyma bands.
- Vessels large in early wood, medium to small in late wood, moderately numerous (4-6/mm^2), solitary or in radial multiples of 2 or 3, round to oval, vessel lines distinct, usually filled with abundant tyloses, clearly visible to the unaided eye.
- Soft tissue forms eyelets around vessels, often join together to form bands.
- Parenchyma paratracheal- predominantly aliform-confluent, abundant, mainly as 'haloes' or eyelets round the vessel or vessel groups. Rays fine, numerous and closely spaced.

Uses: The timber is largely used for construction, door and window frames, battens for tea- chests, packing cases, ammunition and explosive boxes, pent top wooden cases, wooden ladders and trestles, wooden poles for overhead power and telecommunication lines, fence posts, panelled and glazed shutters, furniture and cabinets, tool handles, poles and posts, wooden crates, mine work, ballies, railway sleepers, textile mill accessories, artificial limbs and rehabilitation aids, boat and ship buildings, lorry and bus bodies, clubs, strip flooring, balancing bench, javelins, hurdles, vaulting stands and balancing bars, cooperage, cart and carriages, bent wood articles and toys. Its bending properties are good and makes good bends for common grades of badminton and tennis rackets, boat parts, shafts, walking sticks, etc.

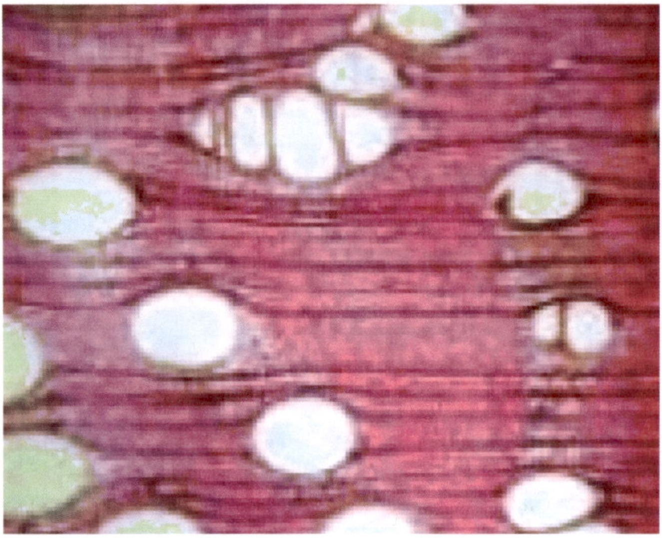

Fig. 13.5. Cross Section of *Lagerstroemia macrocarpa*

6. *Trema orientalis* (Linn.) Blume

Trade Name: Charcoal tree, Indian Nettle tree

Vernacular Names: Sapong, Sempak (Asm.), Chikan (Beng.), Gol (Guj.), Bada Manu (Hind.), Gorklu, Kevuhale (Kan.), Ama, Ayali Malan Totali, Pottama, Rathi, Ambarki (Mal), Kapashi, Kagol (Mar.) Ambarathi, Chenkolam (Tam.), Buda Maru, Chakamaanu (Tel.)

Botanical Name: *Trema orientalis* (Linn.) Blume

Family Name: Ulmaceae

Origin (Distribution) India, Nepal, Bangladesh, Myanmar and Sri Lanka. It occurs in Sub-Himalayan tract from the Jamuna Eastwards to Assam, very scarce to the West, in, central, Western and Southern India, except in the very dry tracts and upto 1500m.

Tree Description: A fast growing but short lived ever green tree or a large shrub, attaining a height of 10 m and a girth of 1m.Bark thin, greenish-grey or bluish-green. It gives a fibre which is used to tie the rafters of native houses and for binding loads.

General features: *Trema orientalis* (Charcoal tree) is a light wood.

- Heartwood off white, light reddish-grey or tinged with pink colour; not distinct from the sapwood.
- **Colour:** Heartwood off white, light reddish-grey or tinged with pink colour; not distinct from the sapwood.
- Weight light; air dry specific gravity approx.0.39
- Grain straight to interlocked.
- Texture fine to medium

Gross Features

- Growth ring boundaries indistinct or absent. Wood is diffuse-porous vessels solitary and outline angular.
- Fibres with simple to minutely bordered pits are present.
- Axial parenchyma absent or extremely rare.
- Axial parenchyma scanty paratracheal.
- Rays are 1 to 3 cells width radial multiples not common, high proportion of solitary vessels. **Uses**: Suitable for manufacturing panel products, poles and drumsticks. Also used in making wooden shoes, fruit boxes, packing, handicrafts, particle board, charcoal, pulp and paper. In Assam it is used to make the coarse Amphak cloth.

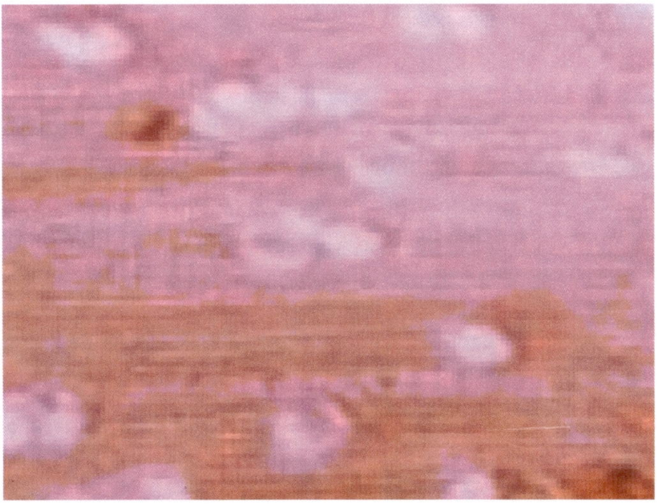

Fig. 13.6: Cross Section of *Trema orientalis*

7. *Grewia tiliifolia* Vahl.

Trade Name: Dhaman

Vernacular Names: Damnak (Bhil.); Dhamani, Daman (Mar.); Dadsal, Butale, Thadsal, Batala (Kan.); Dhamana (Guj.); Dhaman, Baringa (C. P); Khosla, Kasul (Gondi); Charachi, Jana, Tharra, (Tel.); Unu, Chadachchi, Charachi, Unam (Tam.); Pharsa, Phalsa, Dhamin (U. P); Pharsia (Kumaon); Dhaman, Dhamin, Phalsa, Pharsa, (Hind.); Chadache, Chadicha, Sadachu (Mal.); Bhangia, Dhamani, Dhamono and Dhamuro (Or.)

Botanical Name: *Grewia tiliifolia Vahl*

Family Name: Tiliaceae

Origin (Distribution): Southern moist mixed deciduous, teak bearing and semi-evergreen forest. It is a common tree of Himachal Pradesh, Bihar, West Bengal, Assam, Orissa, Rajasthan, Gujarat, Maharashtra, Goa, Andhra Pradesh, Karnataka, Tamil Nadu and Kerala; also in Sri Lanka.

Tree Description: Medium sized tree with 20 m height, 60 cm diameter and 1.5 m in girth with a clear bole of 4-6 m.; bark pale grey in young trees, dark or blackish brown in old trees, fibrous, rough, peels off in thin flakes.

General features: *Grewia tiliifolia* (Dhaman) is a moderately heavy wood.

- Heart wood reddish brown to brown with black or dark streaks.
- Wood is moderately hard to hard, moderately heavy to heavy.
- Parenchyma paratracheal, abundant, vasicentric or aliform.
- Rays fine to broad and close.
- Vessels punctuate with the naked eye, solitary and in radial multiples of two to three; medium-sized; few to moderately numerous (6-10 per mm^2), filled with tyloses or chalky deposits.
- **Colour:** Sapwood yellowish white or pale yellowish, turning brownish grey with age. Heartwood reddish brown to brown with black or dark streaks.
- **Weight:** Moderately hard to hard, moderately heavy to heavy (785 kg m^3 at 12 % m. c.) Grain: Straight Texture Medium to Coarse

Gross features

- Growth rings distinct, demarcated by dense, darker latewood fibers.
- Diffuse porous to semi-ring-porous wood.

- Vessels punctuate with the naked eye, solitary and in radial multiples of two to three; medium to small-sized; few to moderately numerous (6-10 per mm^2), filled with tyloses or chalky deposits.
- Soft tissue forming halos around vessels and also banded; not readily visible, even with a hand lens. Rays fine to broad, close, 5-11 mm; few and spaced. Ripple marks occasionally faintly visible.
- Parenchyma paratracheal, abundant, vasicentric or aliform, visible with a hand lens. Apotracheal parenchyma diffuse and also in tangential lines.

Uses: Agricultural implements, axe-helves, tool handles, constructional purposes like door and window frames, furniture, poles, shafts, panels and bent parts in carriage and spokes in cart construction, ballies, cross arms and fence posts, railway sleepers, boat and ship building, badminton rackets, clubs, balancing bench, hurdles, cricket stumps and bails; lorry and bus bodies and brushwares. It is used by tent- markers for making poles for military tents; it is used for making wagon- bottoms and doors in railway. This timber is using for picker-arms in textile mills, oars, felloes, and as shuttles, rabbet, tubes, pins, cotton tubes and warper bobbins. It is an ornamental timber. Dhaman is specified for handles of pruning knives, slay bottoms and slay cap for cotton looms, beams for gymnastics, temporary nail joined construction, beater heads, anvil blocks, ladders and trestles and fence posts.

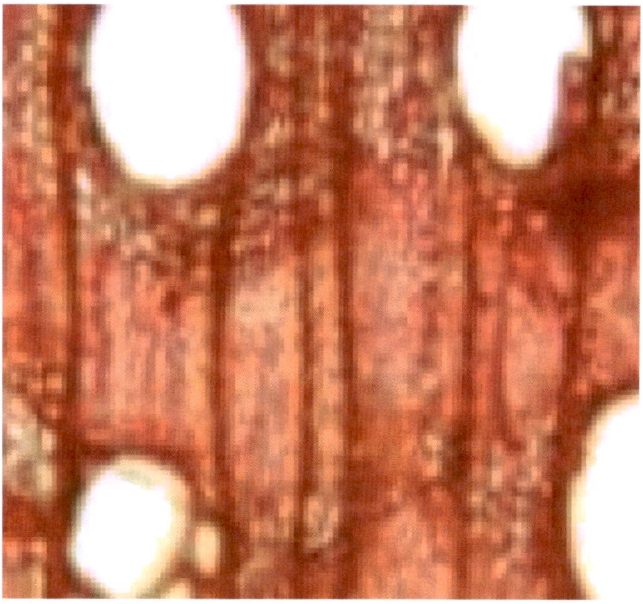

Fig. 13.7. Cross Section of *Grewia tiliifolia*

8. *Diospyros ebenum Koenig ex Retz.*

Trade Name: Ebony Vernacular Names Balemara (Kan.), Karimaram (Mal.), Kendhu (Or.) and Nalluti (Tel.)

Botanical Name: *Diospyros ebenum Koenig ex Retz.*

Family Name: Ebenaceae

Origin (Distribution): Sporadic in West coast semi-evergreen forests. It is also found in the forests of Deccan extending North to the Cudappah and Kurnool districts of Andhra Pradesh, Tamil Nadu and in South Orissa, South Coimbatore, Tirunelvely and Srilanka

Tree Description : Small to medium, occasionally attains a height of 20 m, girth of 2.5 m and about 40 cm in diameter. Bark dark grey with longitudinal fissures. **General features** Diospyros ebenum Koenig. (Ebony) is very hard and very heavy.

- Heartwood is uniform jet black or black brown sometimes with streaks.
- Vessels small, few to moderately few, solitary or in radial multiple of two or three; filled with brownish-black or black gum.
- Parenchyma apotracheal-fine undulating tangential lines.
- Rays very fine and closely spaced. Colour: Heartwood small, uniform jet black or black brown sometimes with streaks; clearly demarcated from the pale red brown or light yellowish grey sapwood.
- Weight: Very hard and very heavy; air dry specific gravity 0.96-1.12 with an average value of 1 (1150 kg/m3 at 12% m.c.).
- Grain: Straight to slightly interlocked or somewhat curly.
- Texture Fine

Gross features

- Wood is diffuse porous.
- Growth rings are indistinct.
- Vessels small, few to moderately few, moderately numerous (6-15 per mm^2) solitary or in radial multiples of two or three, evenly distributed; filled with brownish-black or black gummy deposits in the heartwood.
- Vessel lines indistinct to inconspicuous.
- Parenchyma apotracheal-fine undulating tangential lines, diffuse to diffuse-in-aggregates at places forming a reticulum with the rays.

- Rays very fine closely and evenly spaced.

Uses: It is an attractive timber used for hardwood flooring and inlaid work, parts of musical instruments, handles of cutlery and tools, pocket knives, khukris, walking sticks, toys, decorative carvings, turnery and antiques. It is also used for dobby and pegs in textile mills, mathematical, engineering and drawing instruments, swagger sticks and handicrafts. The sapwood is mainly used for constructional purposes, axe handles, light hammer handles, shafts, shoulder poles, picture frames and furniture.It is a timber for general purpose plywood and for face veneer of commercial type of block boards.

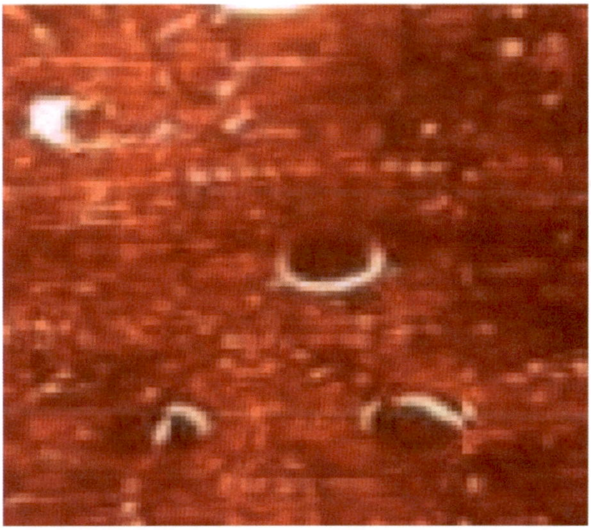

Fig. 13.8. Cross Section of *Diospyros ebenum*

9. *Gmelina arborea Roxb.*

Trade Name: Gamari, Kumbil **Vernacular Names** Gamar (Bangladesh), Gamari, Gumhar, Kumbil, Gomari, Gambar, Shewan, Kumil (India).Yemane (Myanmar, Malaysia, Philippines), Gomari (Asm.),Shewan (Guj.), Gambhar (Hind.),Kulimavu (Kan.), Kumala (Tam.), Gummadi (Tel.)

Botanical Name: *Gmelina arborea Roxb.*

Family Name: Verbenaceae

Origin (Distribution): Native to India, Sri Lanka, Myanmar, Southern China, Laos, Cambodia, Vietnam, Indonesia and introduced to many tropical countries as a fast growing tree species. Sporadic in moist teak bearing and Southern secondary moist mixed deciduous forest.

Tree Description : Medium, 15-20 m in height, 40-65cm in diameter and a girth of 1.5-2 m. Bark whitish grey, corky, warty, lenticellate ex- foliating in regular thin flakes. **General features** *Gmelina arborea* Roxb. (Gamari, Kumbil) is soft to moderately hard and light to moderately heavy.

- Wood is creamy white to pale yellowish-grey turning to yellowish-brown on exposure.
- Parenchyma mostly paratracheal-vasicentric and also in terminal bands delimiting growth rings.
- Rays broad to moderately broad, distinct to the eye, few, rather widely spaced and uniformly distributed.
- Vessels large to medium, distinct to fairly distinct to the eye, rather unevenly distributed, few to moderately numerous, mostly solitary and in short radial multiples of 2 or 3, rounded in outline; tyloses abundant.
- **Colour:** Heartwood and sapwood not distinct in colour; creamy white to pale yellowish- grey or buff turning to yellowish-brown on exposure, moderately lustrous.
- Weight: Soft to moderately hard and light to moderately heavy
- Grain: Straight to interlocked or slightly wavy
- Texture medium to coarse

Gross features

- Wood is diffuse-porous to semi-ring- porous.
- Growth rings scarcely distinct and delimited by comparatively larger early wood vessels and fine lines of parenchyma.
- Vessels large to medium, distinct to fairly distinct to the eye, rather unevenly distributed, few to moderately numerous, mostly solitary and in short radial multiples of 2 or 3, rounded in outline; tyloses abundant.
- Parenchyma mostly paratracheal-vasicentric and also in terminal bands delimiting growth rings. Rays broad to moderately broad, distinct to the eye, few, rather widely spaced and uniformly distributed.

Uses: It is mainly used for building construction, shipbuilding, class I plywood for general purpose, tool handles, artificial limbs and rehabilitation aids, textile mill accessories, cooperage, carrom draughts, tennis and badminton rackets, brushware, musical instruments, shoe-lasts and pencil slats. Highly valued timber for door and window panels, joinery, furniture and cabinets, especially for drawers, wardrobes, cupboards, kitchen and camp furniture because of its light weight, stability and durability. It is also used for bent wood articles, picture frames, toys, block boards, frame core and cross bands of flush door shutters, plane tables, thermometer scales, cheaper grade metric scales, tea-chest, turnery articles, tool handles, instrument boxes, boat building, packing cases and crates, decking and for cars; also used in paper making and matchwood industries.

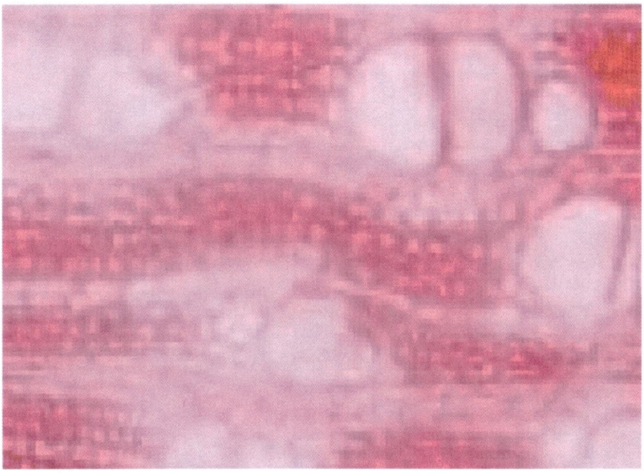

Fig. 13.9: Cross Section of *Gmelina arborea*

10. *Adina cordifolia*

Trade Name: Haldu

Vernacular Names: Manjakadambu, Veembu (Mal), Lampatia, Tarak Chapa (Asm.), Keli- kadam (Beng.), Haladwan, Holdarvo (Guj.), Haldu, Hardu (Hind.), Avanu, Kadambe (Kan.), Manjakadambai (Tam.), Bandaru, Dodaga (Tel.)

Botanical Name: *Adina cordifolia* (Roxb). Hook.f.ex. Brand.(Syn. Haldina cordifolia)

Family Name: Rubiaceae

Origin (Distribution): Southern moist mixed deciduous forests throughout India, except in arid regions of Rajasthan, moist Teak bearing forests and West-coast semi ever green forests. It also occurs in the sub-Himalayan tract from Jamuna Eastwards upto Bhutan and is common throughout central, Western and Southern India, Burma and Sri Lanka.

Tree Description : Medium to very large tree, 15-35 m in height, 2-3.5 in girth with a clear bole of 9 to 12cm and upto 110 cm in diameter; often unevenly fluted and buttressed at the base. Bark grey, soft, thick, exfoliating in small irregular flakes, blaze light red inside with white streaks.

General features

- *Adina cordifolia* (haldu) has heart wood that is deep yellow.
- Heart wood is deep yellow turning reddish or brownish on exposure.
- Wood is light to moderately heavy and has a fine texture.
- Parenchyma is sparse and indistinct to the naked eye
- Rays are fine, numerous and closely spaced.
- Vessels are small, solitary and in radial multiples of two to three
- **Colour:** Sapwood pale yellowish or yellowish white. Heart wood is deep yellow turning reddish or brownish on exposure.
- **Weight:** Moderately hard and light to moderately heavy (695 kg/m^3 at 12% m.c.).
- **Grain:** Fairly straight to somewhat interlocked.
- Texture fine textured

Gross features

- Wood is diffuse porous.
- Growth rings indistinct.
- Vessels solitary and in radial multiples of two to three, small to very small sized, numerous to very numerous (25-45 per mm^2), evenly distributed, oval in outline, open, vessel lines indistinct, not clearly visible to the naked eye.
- Soft tissue indistinct to the eye, but visible as bands under a hand lens.
- Parenchyma extremely sparse paratacheal-scanty; apotracheal, relatively abundant, diffuse and diffuse in aggregate. Rays fine, numerous and closely spaced.

Uses: It is one of the best turnery and carving wood and is largely used for class I ply wood, tea chests, furniture and cabinets, block board and tool handles, pen- holders, combs, reels of cott on and other yarns, jute and textile mill accessories, slubbing and rover tubes and skewers, carrom draughts, chesspieces, bobbins, packing cases and crates, pentop cases, frame, core and cross band of flush door shutters, mine work, ceiling boards, ornamental caskets and picture frames, cricket stumps and bails; musical instruments, mathematical engineering and drawing instruments, brush ware, bent-wood articles and toys; shoe lasts, battery separators, folding chairs and camp furniture and house hold fitments.

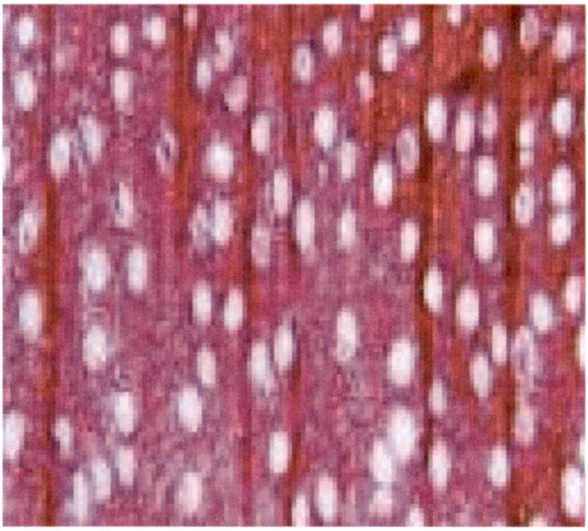

Fig. 13.10: Cross Section of *Adina cordifoli*a

11. *Abies pindrow* Royle

Trade Name: Himalayan Fir

Vernacular Names: Partal, Pindrau, Span,Tos (Hind.), Badon, Drewar (Kash.)

Botanical Name: *Abies pindrow* Royle

Family Name: Pinaceae

Origin (Distribution): Trees are distributed in temperate regions of Northern hemisphere. It occurs from Afghanistan to Nepal between 2300 and 3300 m elevation mostly associated with Deodar, Spruce and Oak.

Tree Description: It is a lofty tree 45-60 m in height and 4 m in girth. Bark is dark grey and rough.

General features

- *Abies pindrow* (Himalayan Fir) is a soft and light wood.

- The wood is creamy white, turning light brown on ageing, without any distinct heartwood and no odour.
- Growth rings are distinct; transition from early to latewood is usually gradual.
- Rays are very fine, distinct only under hand lens as closely-spaced fine lines.
- Resin canals are absent.
- **Colour:** The wood is creamy white, turning light brown on ageing, without any distinct heartwood. No odour
- **Weight:** Soft and light wood. 450 kg/m^3 at 12% m. c.
- **Grain:** Straight texture fine

Gross features

- Growth rings are distinct. Vessels transition from early to latewood is usually gradual.
- Rays are very fine, distinct only under hand lens as closely- spaced fine lines.
- Resin canals are absent

Uses

- It is in great demand for all types of light packing cases and fruit crates.
- Being odourless, it is particularly suitable for packing delicate articles prone to tainting.

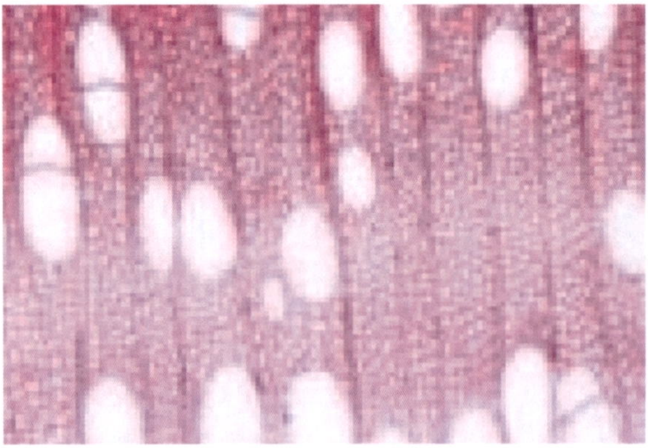

Fig. 13.11: Cross Section of *Abies pindrow*

12. *Anthocephalus chinensis (Lamk.)*

Trade Name: Kadam

Vernacular Names: Kadam (Hin.), Cadamb, Attu –teak, Kodavara, Chakka, Kadambam (Mal.), Roghu (Asm.), Kadam (Beng.), Kadamb, Karam (Hind.), Bale (Kan.), Kadamba, Nhyu (Mar.), Kadambo (Or.), Vellei- Kadamba, Kola-aiyila (Tam.), Kadambe (Tel.)

Botanical Name: *Anthocephalus chinensis* (Lamk.) A.Rich. ex Walp. and *Anthocephalus cadamba* (Roxb.) Miq.

Family Name: Rubiaceae

Origin (Distribution): West coast semi-evergreen forests. It occurs in the sub-Himalayan tract from Nepal Eastwards in the lower hills of Darjeeling, Assam, Bihar, Orissa, Eastern Ghats, Andamans and Burma. It is a tree of moist warm regions, deciduous and evergreen forests.

Tree Description : Medium to large tree with a straight cylindrical bole and horizontal branches, 15-25 m in height and 1.5-2.5 m in girth with a clear bole of 9 m and about 60 cm in diameter. Bark dark grey with longitudinal fissures; peels off in thin scales.

General features

- Wood of *Anthocephalus chinensis* is white with a yellowish tinge and heartwood that is not distinguishable from sap wood.
- Wood is light to moderately heavy.
- Growth rings are fairly distinguishable.
- Parenchyma is scanty.
- Rays are coarse-storied.
- Vessels are large to very small and moderately numerous to numerous, mostly solitary or in radial multiples of 2, 3 or 4.
- **Colour:** Heart wood and sapwood are not distinct. White with yellowish tinge to cream white or yellowish grey.
- **Weight:** Light to moderately heavy, 385 to 640 kg/m^3 at 12 % m. c.
- **Grain:** Straight texture medium to fine

Gross features

- Wood is diffuse porous.
- Growth rings fairly distinct, but not conspicuous.
- Vessels large to very small, moderately numerous to numerous (4-15 per mm^2), mostly solitary or in radial multiples of 2, 3 or 4,oval in outline, open, vessel lines distinct.
- Parenchyma paratracheal-scanty; apotracheal diffuse to diffuse-in-aggregates, tending to form a net like structure with rays. Rays very fine, closely spaced.

Uses: It is extensively used for packing cases, ceiling boards, bottom of drawers, backing of cupboards, toys. It is also used for furniture, tea chests, building construction, class III veneers for plywood, match boxes and match splints, soles and heels of sandals and shoes, pencil slats, turnery and battens. It is also suitable for writing and printing paper.

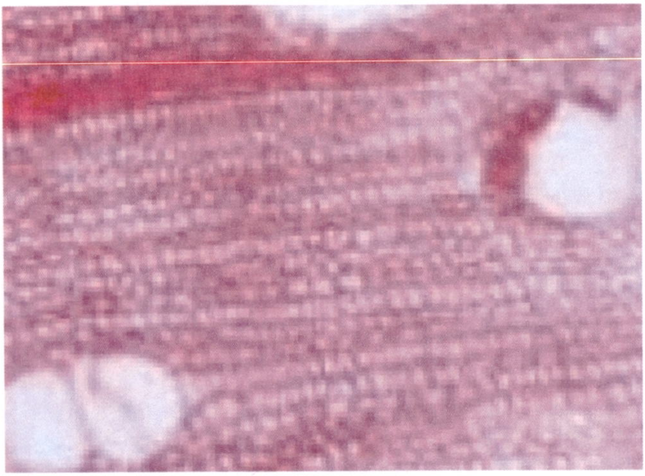

Fig. 13.12: Cross Section of *Anthocephalus chinensis*

13. *Holoptelea integrifolia (Roxb.)*

Trade Name: Kanju, Indian Elm

Vernacular Names: Anjan (Beng.); Chirol, Karanji, Papri (Hind.); Aval (Mal.); Papara (Mar.); Thapsi (Kan.); Chilbil (Or.); Ayili (Tam.)

Botanical Name: *Holoptelea integrifolia (Roxb.) Planch.*

Family Name: Ulmaceae

Origin (Distribution): West coast semi-evergreen and occasionally in moist Teak bearing forest.

Tree Description: Medium to large-sized, 15-25 m in height and 80 cm in diameter; bark whitish-grey, thin, with longitudinal furrows and exfoliating in irregular flakes

General features

- *Holoptelea integrifolia* (Kanju) is a moderately heavy wood.
- Heartwood and sapwood are not distinct.
- Wood light yellow or yellowish grey.
- Wood is moderately hard and moderately heavy.
- Parenchyma paratracheal, aliform and winged type.
- Rays fine to moderately broad, closely spaced and numerous.
- Vessels solitary and in radial multiples of two to three, medium- sized
- **Colour:** Heartwood and sapwood are not distinct.Wood light yellow or yellowish grey, darkening on exposure; somewhat lustrous.
- **Weight:** Moderately hard and moderately heavy (595-600 kg/m^3 at 12 % m.c.)
- **Grain:** Shallowly interlocked texture medium coarse

Gross features

- Growth rings distinct.
- Diffuse porous wood.Vessels solitary and in radial multiples of two to three, medium-sized, moderately numerous (5-8 per mm^2), filled with white chalky deposits.
- Soft tissue forms eyelet shaped patches around vessels.
- Rays fine to moderately broad, closely spaced and numerous. Ripple marks present.
- Parenchyma paratracheal, aliform and winged type; also confluent bands of one to three cell widths present.

Uses

The wood is used for constructional works, bobbins and cotton reels in textile mills, general purposes class I plywood, tea chests, decorative plywood, block boards, furniture, and cabinets, tool handles, agricultural implements, bentwood articles and toys.

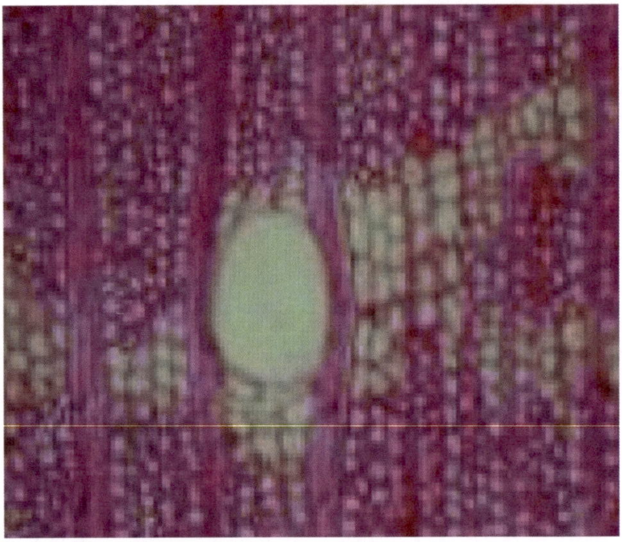

Fig. 13.13: Cross Section of *Holoptelea integrifolia*

14. *Prosopis juliflora* DC

Trade Name: Keekar

Vernacular Names: Perdesi Baval (Guj.), Vilayati Babul, Vilayati Khejra (Hindi), Vilayati Jhand, Vilayati Kikar (Punjab & Haryana), Ganda Babul, Ghaf Bahri or Velvet Mesquite (India)

Botanical Name: *Prosopis juliflora* DC

Family Name: Leguminosae (Mimosae)

Origin (Distribution): A native of tropical America introduced and now naturalized throughout arid areas in India chiefly Andhra Pradesh, Delhi, Haryana, Karnataka, Madhya Pradesh, Maharashtra, Punjab, Rajasthan, Tamil Nadu and Uttar Pradesh. It is usually seen in ever green and semi-ever green forests.

Tree Description: It is a small to medium-sized tree usually unarmed, attaining a height of 9-12 m and girth of 90 cm. It may attain a height of 18 m in favourable localities. Bark greyishbrown, rough, longitudinally fissured and thin.

General features

- *Prosopis juliflora* DC (Keekar) is a hard to very hard wood.
- Heartwood dark brown to purplish-brown.
- Wood is very hard.
- Parenchyma paratracheal, mostly aliform to aliform confluent, enclosing several vessels at places forming wavy band particularly in the late wood also as diffuse scattered cells and terminal/ marginal in broken lines on growth rings, strands fusiform.
- Rays homogenous, composed of procumbent cells only, gummy infiltrations common.
- Vessels moderately large to small, visible to the eye, 45-190 μm in diameter (mean 115 μm, few to moderately numerous or numerous (13-26/ mm^2).
- The mesquite pods are sweet and edible and flowers are the source of honey.
- **Colour:** Sapwood distinct from heartwood. Heartwood dark brown to purplish-brown and sapwood yellowish –white or buff
- **Weight:** Hard to very hard and heavy to very heavy.
- **Specific gravity** 0.88- 0.90 (Air- dry)
- **Grain:** Straight to slightly interlocked texture medium to coarse

Gross features

- Wood is diffuse porous.
- Growth rings inconspicuous but usually distinct in the sapwood under the hand lens, delimited by a fine interrupted line of parenchyma and denser zone of latewood fibres, 1-5 per cm.
- Vessels moderately large to small, visible to the eye, 45-190 μm in diameter (mean 115 μm, few to moderately numerous or numerous (13-26/mm^2).

- Parenchyma paratracheal, mostly aliform to aliform confluent enclosing several vessels at places forming wavy band particularly in the late wood also as diffuse scattered cells and terminal/marginal in broken lines on growth rings, strands fusiform.
- Fibres libriform, angular in cross section, sometimes arranged radially, non- septate. Rays homogenous-composed of procumbent cells only, gummy infiltrations common.

Uses: The mesquite pods are sweet and edible and flowers are the source of honey. The gum obtained is used as an adulterant or substitute for gum arabic. The wood takes paint and polish well. It is cheaply valued for fuel and charcoal but locally can be used for tool handles, agricultural implements and cheap furniture.

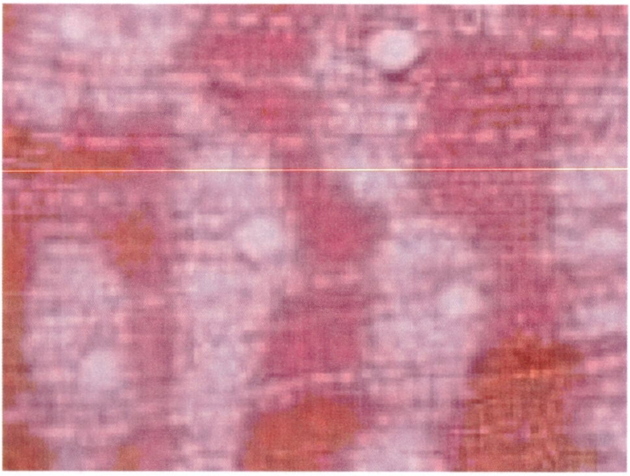

Fig. 13.14: Cross Section of *Prosopis juliflora*

15. *Swietenia macrophylla*

Trade Name: Mahogany

Vernacular Names: Honduras Mahogany, Guatemala Mahogany, Brazilian Mahogany, Mogno (Brazil), Zopilote and Chiculte (Mexico)

Botanical Name: *Swietenia macrophylla* King

Family Name: Meliaceae

Origin (Distribution): Native to Central and South America, particularly Mexico and Honduras, and introduced to many tropical countries including India. It is widely cultivated as an avenue tree.

Tree Description: A medium sized tree reaching a height of 12-18 m and a diameter of 80-130 cm

General features: *Swietenia macrophylla* (bigleaf mahogany) is a hardwood.

- Heartwood is reddish brown.
- Wood is moderately hard and heavy.
- Prominent marginal parenchyma.
- Rays are coarse-storied.
- Vessels are large (easily visible with naked eye), often in multiples of 2/3. It is a CITES Appendix II timber species.
- **Colour:** Heartwood colour varies from light reddish or yellowish brown to dark reddish brown, lustrous. Sapwood yellowish white to pale brownish gray.
- Weight: Moderately heavy 650 kg/m^3 at 12% m. c.
- Grain: Straight to interlocked. Flat-sawn surface produce prominent growth ring figure.
- Texture medium to coarse

Gross features

- Wood is diffuse porous.
- Growth rings distinct.
- Vessels solitary and in radial multiples of two to three, medium-sized, moderately numerous (10-15 per mm^2).
- Soft tissue banded and forms incomplete sheaths around vessels.
- Pores usually stained or discoloured, darker, are large and evenly spread through the wood.
- Rays fine to moderately broad, few and widely spaced. Rays can be seen clearly on radial-cut surfaces as smooth plates which reflect the light.

Uses: High class furniture and cabinet making, paneling and interior joinery, boat interiors, musical instruments, show cases, counters and interior decoration, jewellery boxes, carvings, rotary cut logs for ply wood and sliced veneers for decorative work.

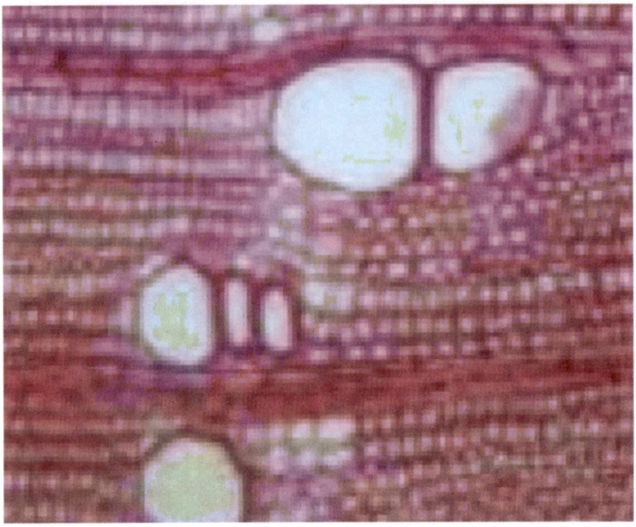

Fig. 13.15. Cross Section of *Swietenia macrophylla*

16. *Acacia mangium Willd.*

Trade Name: Mangium

Vernacular Names: Brown Salwood, Black Wattle, Hickory Wattle (Aus.); Mangge Hutan, Tongke Hutan (Ceram), Nak (Maluku), Laj (Aru), Jerri (Irian Jaya) (Indonesia); Arr (Papua New Guinea); Mangium, Kayu Safoda (Malaysia) and Kra Thin Tepa (Thailand)

Botanical Name: *Acacia mangium Willd.*

Family Name: Leguminosae (Fabaceae)

Origin (Distribution): An exotic, native to Australia, Papua New Guinea and Indonesia; widely cultivated in different parts of Kerala.

Tree Description: A medium-sized tree, reaching a height of 10-18 m and a diameter of 60-70 cm.

General features

- *Acacia mangium* is moderately heavy wood.
- Heartwood is yellowish brown.
- Wood is moderately hard and moderately heavy.
- Parenchyma paratracheal and vasicentric.
- Rays fine, numerous and closely spaced.

- Vessels solitary and in radial multiples of two to three.
- **Colour:** Heartwood is yellowish brown and sapwood is creamy white in colour.
- **Weight:** Moderately hard and moderately heavy (70 kg/m^3 at 12 % m.c.)
- **Grain:** Straight to interlocked
- **Texture** Medium to fine

Gross features

- Growth rings fairly distinct.
- Diffuse porous wood.
- Vessels solitary and in radial multiples of two to three; large to medium-sized; moderately numerous (10-16 per mm^2).
- Soft tissue forming a sheath around vessels.
- Rays fine, numerous and closely spaced.
- Rays one to three-cell wide, homogeneous.
- Parenchyma paratracheal and vasicentric

Uses: Used for furniture making, agricultural implements and construction work. Mainly used for pulp wood production. Suitable for door and window frames, cabinet making, light structural work, moldings, boxes, crates, paneling and turnery, sports goods and charcoal. It is used as a component of composite wood products.

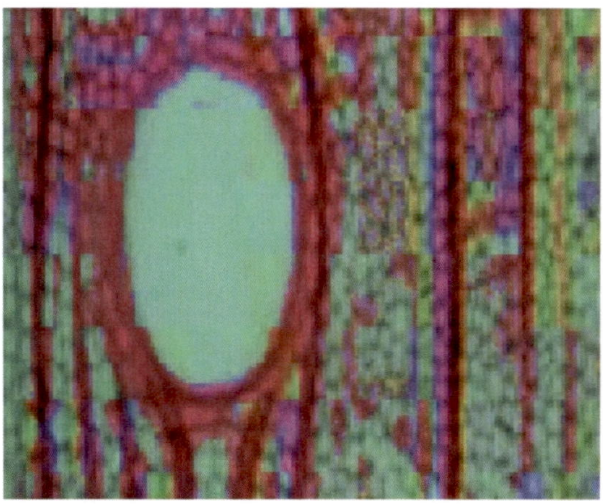

13.16. Cross Section of *Acacia mangium*

17. *Mangifera indica*

Trade Name: Mango wood

Vernacular Names: Maavu, Moochi (Mal.) and Aaam (Hin.)

Botanical Name: *Mangifera indica* Linn.

Family Name: Anacardiaceae

Origin (Distribution): West coast tropical evergreen and West coast semi evergreen forests; cultivated extensively.

Tree Description: Medium to large 15-30 m in height and 50-100 cm in diameter. Bark brown or dark grey, rough.

General features

- *Mangifera indica* (mango wood) is a moderately hard and heavy wood.
- It has yellowish white to greyish brown and somewhat lustrous wood.
- Growth rings are fairly distinct.
- Wood is moderately hard and heavy.
- Aliform to confluent parenchyma and often delimiting growth rings.
- Rays are fine to moderately broad and numerous.
- Vessels are large to medium, often filled with tyloses, numerous and solitary or in radial multiples of 2-3.
- **Colour:** Yellowish white to greyish brown, sap wood and heart wood not distinct or sometimes heartwood distinct and dark brown; somewhat lustrous.
- **Weight:** Moderately hard and moderately heavy. 690 kg/m^3 at 12% m.c
- **Grain:** Straight to somewhat interlocked
- **Texture**: Medium to coarse.

Gross features

- Wood is diffuse porous. Growth rings fairly distinct.
- Vessels large to medium, few to moderately numerous solitary or in radial multiples of 2-3 or more, often filled with tyloses.

- Parenchyma paratracheal- aliform to confluent, often delimiting growth rings.
- Rays fine to moderately broad, numerous, closely spaced.
- Pith flecks are usually present.

Uses

Ceiling boards, window frames, general purpose class I plywood, furniture and cabinets, block boards, match splints and boxes, boat and ship buildings, bobbins, bent wood articles, shoe-lasts.

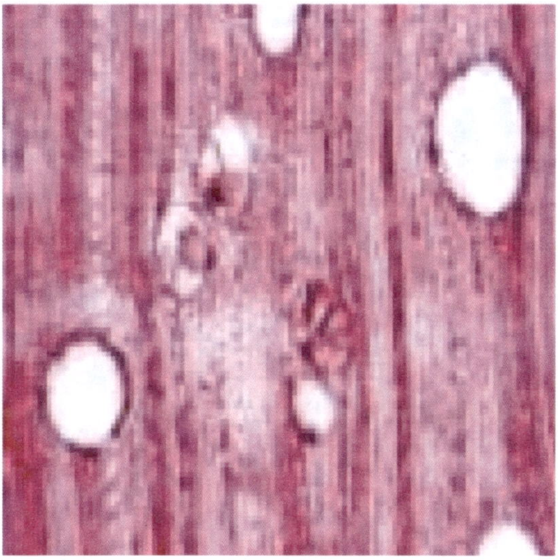

Fig. 13.17. Cross Section of *Mangifera indica*

18. *Eucalyptus tereticornis*

Trade Name: Mysore gum, *Eucalyptus* hybrid

Vernacular Names: *Eucalyptus*, Eucali (Mal.), Mysore gum, Forest red gum, *Eucalyptus* (India), Queensland blue gum (Australia)

Botanical Name: *Eucalyptus tereticornis* Sm.

Family Name: Myrtaceae

Origin (Distribution): Native of Australia extensively raised in plantations in many tropical and sub tropical countries including India for pulpwood production. Mostly seen in Uttar Pradesh, Madhya Pradesh, Karnataka, Tamil

Nadu and West Bengal. It is also seen in Long Islands in Middle Andaman Division. It is indigenous to coastal districts of Eastern Australia and Papua New Guinea. In India it is the most widely planted species upto 1,200 m elevation.

Tree Description

Large, about 25-45 m in height, 3-6 m in girth, and straight bole of about 15-22 m and large open crown and 40 cm in diameter. Bark smooth, whitish or ash coloured, deciduous decorticating in more or less long plates and irregularly blotched throughout, sometimes with a little rough and flaky bark persisting at base.

General features

- *Eucalyptus tereticornis* Sm. (Mysore gum) is a very hard and heavy wood.
- Heartwood is pale brown to reddish brown.
- Parenchyma: moderately abundant to abundant, apotracheal, diffuse.
- Rays fine, closely and more or less evenly spaced.
- Vessels medium sized, visible to the eye, moderately numerous (5-20 per mm^2), mostly solitary, open or filled with abundant tyloses, kino-like deposits sparse to moderately abundant.
- **Colour:** Heart wood pale brown to reddish brown fairly well demarcated form the sapwood. Sapwood light greyish-brown or pale red.
- **Weight:** Lustrous, hard to very hard. Heavy to very heavy. Air dry specific gravity 980 kg/m^3
- Grain: Straight to shallowly interlocked or wavy.
- Texture medium to coarse.

Gross features

- Wood is diffuse porous.
- Growth rings are indistinct.
- Vessels medium, visible to the eye, moderately numerous(5-20 per mm2), mostly solitary, rarely in radial or oblique chains, round to oval, open or filled with abundant tyloses, kino-like deposits sparse to moderately abundant.

- Parenchyma moderately abundant to abundant, apotracheal-diffuse to diffuse- in aggregates, kino-like deposits sparse to abundant.
- Parenchyma indistinct to distinct to the eye, distinct under hand lens forming thin sheet round the vessel.
- Rays fine, closely and more or less evenly spaced, kino-like deposits abundant. Specific gravity of this species varies in different localities.

Uses: Mainly used for fuel, paper and pulping, suitable for packing cases and boxes, beams, columns, power transmission poles and posts. It is now being used by some paper mills in North India in mixture with other raw materials. It is also used for rayon grade pulp by some mills. It is highly suitable for construction work, furniture and stakes.

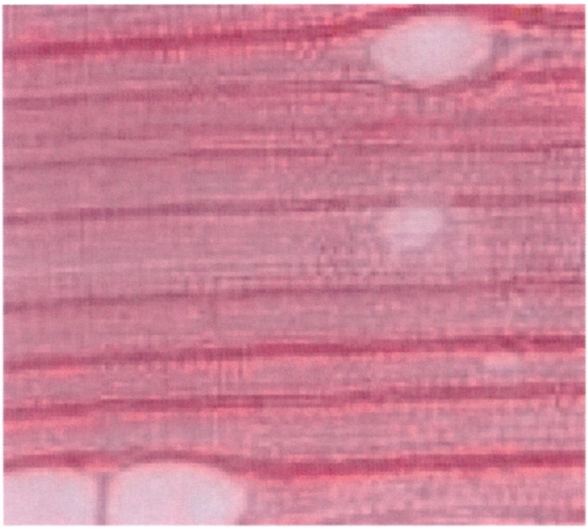

Fig. 13.18: Cross Section of *Eucalyptus tereticornis*

19. *Azadirachta indica*

Trade Name: Neem Vernacular Names Aryaveppu, Veppu, Vempu, Leemdo, Kadunimb, Neem (India), Nim (Pakistan), Baypay (Malaysia) Kwinin (Thailand) , Mindi (Indonesia)

Botanical Name: *Azadirachta indica* A. Juss

Family Name: Meliaceae

Origin (Distribution): Native of Indian subcontinent; distributed throughout South- East Asia, East and SubSaharan Africa, Fiji and some parts of Central America. Naturally found in deciduous forests of Peninsular India and homesteads of Kerala.

Tree Description: Medium to large tree with a height of 15-20 m with a clear bole of 7 m and diameter of 50 cm.

General features

- *Azadirachta indica* A. Juss (Neem) is a moderately heavy wood.
- Heartwood is reddish brown, aromatic and slightly lustrous.
- Distinct growth rings
- Parenchyma: Apotracheal, paratracheal- vasicentric, also in tangential lines connecting vessels.
- Vessels medium sized, few and occurring in solitary or in radial multiples of two to three, often in clusters and filled with reddish gummy deposits.
- **Colour:** Heartwood reddish brown, aromatic, moderately lustrous. Sapwood is yellowish white, yellowish brown or greyish yellow.
- **Weight:** Hard to moderately hard; moderately heavy. (835 kg/m^3 at 12% m.c.)
- **Grain:** Interlocked texture coarse strength.

Gross features

- Growth rings distinct.
- Vessels medium, few to moderately few, solitary or in radial multiples of two to three, often in clusters; numerous (16-20 per mm^2).
- Vessels filled with reddish gummy deposits.
- Soft tissue forms bands delimiting growth ring and also associated with vessels.
- Parenchyma – Apotracheal-irregularly placed tangential and continuous bands delimiting growth rings, Paratracheal- Vasicentric, also in tangential lines connecting vessels.
- Rays fine to medium sized, numerous and somewhat widely spaced.

Special Features

- The wood possesses characteristic odor.
- The wood has insect repellant properties due to the presence of neem oil.
- Gum canals often present in tangential bands.

Uses: Used in light construction, furniture, doors and window frames boards and panels, cabinets, boxes and crates. Also used for agricultural implements, tool handles, musical instruments, cigar boxes, matches, ply wood, veneers, carving, and toys.

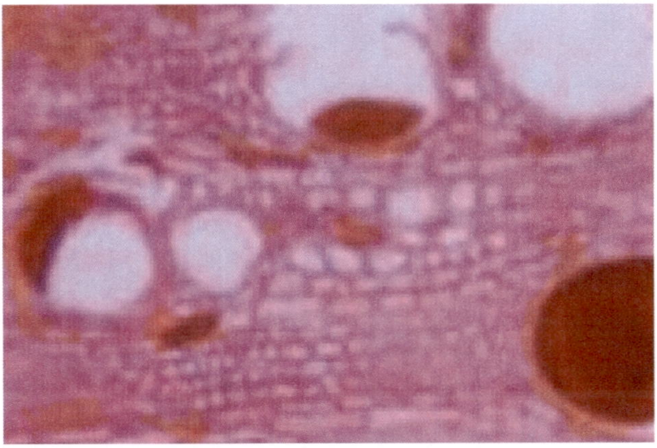

Fig. 13.19. Cross Section of *Azadirachta indica*

20. *Terminalia arjuna*

Trade Name: Arjun

Vernacular Names: Arjun (Beng., Hind., Or.), Arjuna- sadada (Guj.), Holematti, Nadiam (Kan.),Kahu, Sandura (Mar.), Kula muruthu, Vella marda (Tam.) andYermaddi (Tel.)

Botanical Name: *Terminalia arjuna* Bedd.

Family Name: Combretaceae

Origin (Distribution): The tree occurs throughout greater part of the Peninsula from Avadh Southwards on banks of streams. It is found in Bihar, Orissa, Coastal Andhra Pradesh, the Deccan and Central India, but uncommon in Tamil Nadu except in Tirunelveli. It occurs on the West Coast from Gujarat Southwards. It also occurs locally along banks of streams in sub-Himalayan tract in Uttar Pradesh and Himachal Pradesh.The tree is extensively planted for shade and ornament in avenues and parks.

Tree Description: Large evergreen tree with spreading crown and drooping branchlets, 18-24 high and over 3 m in girth. The bole is rarely long or straight and is usually buttressed and often fluted. Bark is pale greenish-grey and smooth.

General features

- *Terminalia arjuna* is a hard and heavy wood.
- Sapwood is reddish-white and the heartwood, which is sharply delineated from the sapwood, is brown to dark brown variegated with darker coloured streaks.
- Wood is diffuse-porous.
- Growth rings delimited by fine concentric lines of soft tissues which are distinct only under hand lens.
- Vessels - The pores are large, distinct to the eye, few, more or less evenly distributed solitary or in radial multiples of 2-3, partially filled with tyloses and also reddish-brown gummy deposits in the heartwood.
- Soft tissues form sheath or eye lets round the pores which are usually distinct to the eye, occasionally also with short lateral extensions which connect adjoining pores.
- Rays are fine to very fine, distinct only under hand lens.
- **Colour:** Sapwood is reddish-white and the heartwood, which is sharply delineated from the sapwood, is brown to dark brown variegated with darker coloured streaks.
- Weight: Hard and heavy. Air-dry weight about 800 kg/cum
- **Grain:** Interlocked
- **Texture:** Coarse

Gross features

- Wood is diffuse-porous.
- Growth rings delimited by fine concentric lines of soft tissues which are distinct only under hand lens.
- Vessels - The pores are large, distinct to the eye, few, more or less evenly distributed solitary or in radial multiples of 2-3, partially filled with tyloses and also reddish-brown gummy deposits in the heartwood.
- Soft tissues form sheath or eye lets round the pores which are usually distinct to the eye, occasionally also with short lateral extensions which connect adjoining pores.
- Rays are fine to very fine, distinct only under hand lens.

Uses: The timber is mainly used for agricultural implements, water troughs; boat building, cart making and pit props. It is also used for constructional purposes like door and window frames. It is also used for block boards and plywood. As the trees are found on river banks, they are often not felled for fear of erosion.

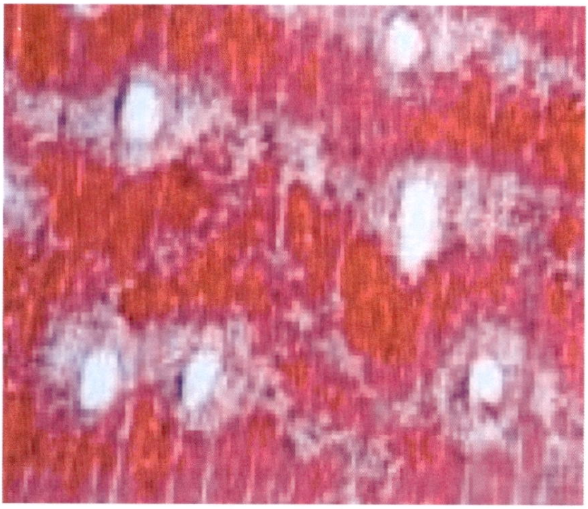

Fig. 13.20: Cross Section of *Terminalia arjuna*

21. *Bombax ceiba*

Trade Name: Semul

Vernacular Names: Himila (Asm.), Simal(Beng.), Khatsaweri (Bhil.), Shimal (Garhwal), Bolchu (Garo.), Bargu (Gon.), Sawar (Guj.), Semul, Semur (Hind.), Bonju-phang (Kach), Shimlo (Kamrup), Burla (Kan.), Savar (Mar.), Simbal (Punj.), Illavam (Tam.), Buraga (Tel.), Mullilavu, Poola and Cotton tree (Mal.)

Botanical Name: *Bombax ceiba* L.Syn and *Bombax malabaricum* (DC.) *Salmalia malabarica* (DC.) Schott. and Endl.

Family Name: Bombacaceae

Origin (Distribution): Widely distributed throughout India, except in very arid tracts, ascending to over 1,500 m. Commonly it is seen in West coast semi evergreen, Southern moist mixed deciduous and moist Teak bearing forests. Occasionally it tends to be gregarious on alluvial soils near river banks, also occur in Myanmar.

Tree Description: Large to very large, deciduous 25-40 m in height with a clear bole of 15-25 m and upto 150cm in diameter, 3 m in girth; huge buttressed 4.5-6 m in height. Under favourable condition, it grows to enormous sized trees 59 m in height and 4.5 m in girth. On poor soil and higher elevations, it is usually stunted. Bark grey, covered with conical prickles when young, deeply cracked when old, outer bark fleshy and soft, inner fibrous, upto 2.5 cm in thickness.

General features

- *Bombax ceiba* is a light wood.
- Heart wood and sap wood are not distinct.
- Creamy white to pale yellowish-brown or yellowish-brown, greyish brown on exposure, often lustrous showing silver effect on radial surface.
- Diffuse porous wood.
- Growth rings not always distinct to the eye, but clearly visible under hand lens, delimited by the slightly denser latewood and closely spaced bands of parenchyma.
- Parenchyma predominantly apotracheal-visible only under the lens, diffuse to diffuse-in-aggregate, closely spaced, fine interrupted tangential line forming reticulum with rays.
- Vessels very large to large, clearly visible to the eye, very few to few, scanty, mostly solitary or in radial multiples of 2 or 3,oval in outline, usually open or partly filled with tyloses.
- Rays fine to very broad, widely spaced, forming conspicuous flecks on radial surface, distinct only under lens, interspersed with broader rays sometimes tending to be storied. Ripple marks usually present.
- **Colour:** Heart wood and sap wood are not distinct. Creamy white to pale yellowish-brown or yellowish-brown, greyish brown on exposure, often lustrous showing silver effect on radial surface.
- **Weight:** Very light to light (365kg/m^3 at 12% m.c.)
- **Grain:** Straight texture coarse

Gross features

- Wood is diffuse porous .
- Growth rings not always distinct to the eye, but clearly visible under hand lens, delimited by the slightly denser latewood and closely spaced bands of parenchyma, usually less than 2 per cm.

- Vessels very large to large, clearly visible to the eye, very few to few, usually more or less evenly distributed, scanty, mostly solitary or in radial multiples of 2 or 3,oval in outline, usually open or partly filled with tyloses, vessel lines conspicuous on the longitudinal surfaces, vessel elements being individually distinct 20-25 per cm.
- Parenchyma predominantly apotracheal-visible only under the lens diffuse to diffuse in-aggregate, closely spaced, fine interrupted tangential line forming reticulum with rays.
- Lines of parenchyma same or wider than fibre layers, 10-15 per mm.
- Rays fine to very broad, widely spaced,forming conspicuous flecks on radial surface, just visible or indistinct to the eye, distinct only under lens, interspersed with broader rays sometimes tending to be storied.
- Ripple marks usually present, distinct to just visible to the eye, but sometimes under the lens, 20-22 per cm due to storied parenchyma.

Uses: It is used for making class III plywood and veneers, packing cases and boxes, match splints and boxes, ship building, fishing floats, cooperage toys, pencil slates, dug-outs, drums and cheap grade pencils, returnable wooden crates, shooks of wooden packaging. 'Compreg' from semul has proved suitable for bosshead of air screw blades.

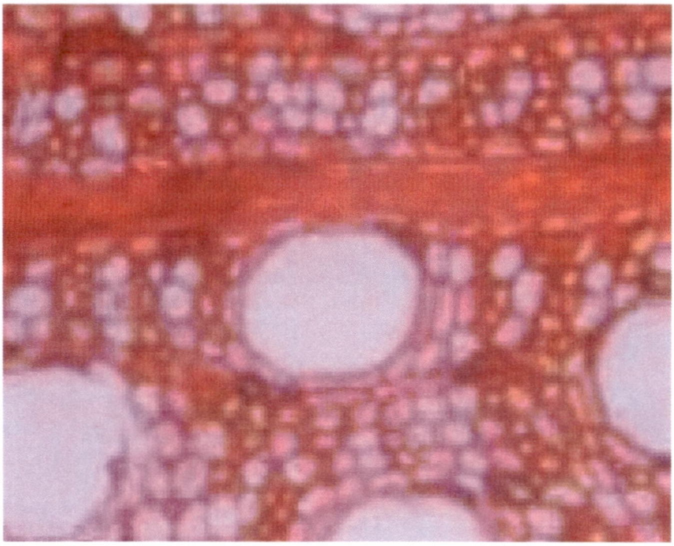

Fig. 13.21: Cross Section of *Bombax ceiba*

22. *Dalbergia sissoo* Roxb.

Trade Name: Sissoo

Vernacular Names: Shisham, Sissoo (India), Sissou (Nepal), Hihu (Asm.), Sisam, Shishma (Guj.), Agara, Birdi (Kan.), Shishai (Punj.), Shinshapa (San.), Gette, Yette (Tam.), Sinsupa (Tel.)

Botanical Name: *Dalbergia sissoo* Roxb.

Family Name: Leguminosae (Fabaceae)

Origin (Distribution): Punjab to Assam in the Sub-Himalayan tract in India and Nepal, Bangladesh, Bhutan, Myanmar, Afghanistan, Pakistan, Iran and Iraq.

Tree Description: Small to large up to 30 m in height with curved or crooked and buttressed stem. Bark grey or brown, longitudinally and somewhat reticulately furrowed, thick.

General features

- Heartwood of *Dalbergia sissoo* (Shisham) is golden brown to dark brown with deep dark streaks.
- Hard and moderately heavy wood.
- Growth rings are distinct to indistinct, demarcated by a fine line of parenchyma.
- Parenchyma is aliform to aliform confluent, also forming straight to wavy narrow bands in late wood.
- Vessels are large and few in early wood and small and few in late wood, often filled with dark, gummy deposits, solitary and in radial multiples of 2 or 3.
- **Colour:** Heartwood golden brown to dark brown with deep dark streaks, soon becoming dull, clearly demarcated from the pale brownish to white sapwood.
- **Weight:** Hard and moderately heavy; 820 kg/m^3 at 12 % m. c.
- **Grain:** Narrowly interlocked.
- Texture medium to coarse

Gross features

- Wood is diffuse porous.
- Growth rings are distinct to indistinct, demarcated by a fine line of parenchyma but indistinct or inconspicuous.
- Vessels large to small, few to moderately few, rather unevenly distributed, generally larger and comparatively more numerous in early wood, mostly solitary and in radial multiples, open but occasionally filled with white or dark, gummy deposits, perforation simple, inter vessel pits vestured and small to medium sized.
- Parenchyma aliform to aliform confluent also forming straight to wavy narrow bands in late wood. Terminal or marginal, delimiting growth rings, strands are storied, fusiform often subdivided into crystalliferous locules.
- Fibres libriform, round to angular in cross section, often gelatinous, non-septate, storied, inter fibre pits small, simple to narrow bordered.
- Rays mostly uniseriate but may be up to 2 or 3 ceriate, 7-16/mm, homogenous weakly heterogenous composed of mainly of procumbent cells.

Uses: First class timber for cabinetry and furniture, paneling and flooring. It yields a very strong veneer cut on a rotary machine which is also highly decorative. Also used for boards, rafters, posts, boat building, truck and lorry bodies, door, window shutters and frames, carvings and engraving, printing blocks, brush backs, superior walking sticks, umbrella handles, pencil slates, sport goods, mallet heads, anvil blocks flooring and ceiling furniture, agricultural and musical instruments and tool handles. It is used for marine and air craft grade plywood, decorative plywood and block boards. The root wood is suitable for tobacco pipes.

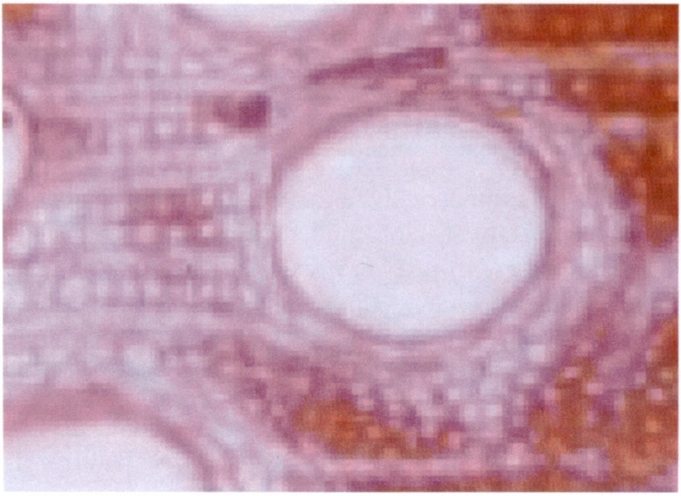

Fig. 13.22: Cross Section of *Dalbergia sissoo*

23. *Grevillea robusta*

Trade Name: Silver oak

Vernacular Names: Southern Silky-Oak (Australia), Lacewood (USA), Silky-Oak (Indonesia) and Savukkumaram (Tam.),

Botanical Name: *Grevillea robusta* A. Cunn. Ex R. Br.

Family Name: Proteaceae

Origin (Distribution): Native to Eastern Malaysia, New Caledonia and Australia, extensively planted as a shade tree for coffee and tea plantations in Africa, India, Sri Lanka and other parts of the world.

Tree Description: An evergreen tree attaining a fair height with long conical crown. Medium to large 18-25 m in height and about 65 cm in diameter. Bark dark grey and rough with vertical fissures.

General features

- *Grevillea robusta* A. Cunn. Ex R. Br. (Silver Oak) is a hard and moderately heavy timber.
- Wood is pinkish brown becoming yellow brown on exposure.
- Parenchyma paratracheal- vasicentric and aliform to aliform- confluent, just visible to the naked eye.
- Rays are of two types; broad to very broad and fine to very fine.

- Vessels large to medium, moderately numerous (3-9/ mm^2), mostly in tangential rows filled with deposits and separated by fibrous tissue in alternate fashion, usually in irregular clusters or short radial or tangential multiples, rarely solitary, round to oval, mostly open and vessel lines distinct.
- **Colour:** Heartwood distinctly pinkish brown becoming yellow brown on exposure, lustrous, well demarcated from the cream coloured sapwood.
- **Weight:** Hard and moderately heavy
- **Grain:** Straight to wavy
- **Texture:** Medium to coarse

Gross features

- Wood is diffuse porous.
- Growth rings are fairly distinct.
- Vessels large to medium, moderately numerous (3-9/mm^2), mostly in tangential rows filled with deposits and separated by fibrous tissue in alternate fashion, usually in irregular clusters or short radial or tangential multiples, rarely solitary, round to oval, mostly open and vessel lines distinct.
- Parenchyma paratracheal- vasicentric and aliform to aliform-confluent, just visible to the naked eye, distinct under hand lens in closely spaced lines enclosing or connecting the vessels tangentially (18-34/ cm).
- Rays are of two types; broad to very broad and fine to very fine, the former prominent to the naked eye, widely spaced and the later rarely visible under the hand lens.
- Prominent silvery radial flecks present.
- Gum canals of traumatic type occasionally present in tangential rows.

Uses: Used for carpentry, cabinet work, joinery, furniture, parquet flooring, interior finishes, paneling, turnery and light construction work. Also suitable for bent wood furniture, packing cases and crates, boxes, truck bodies, naval uses and handicrafts. It is used for building construction, door shutters, class III general plywoods, boards, mathematical engineering and drawing instruments, high speed jute bobbins and rehabilitation aids. It is a highly decorative timber and is used for making decorative plywood and veneers, core and cross bands of block boards and flush doors. Also used to make spool centres for jute spool winding machines, wedges of tennis rackets, brush backs and slate frames. In some countries, it is used for furniture in place of Oak and also for flooring in the form of blocks and strips as its resistance to abrasion is high.

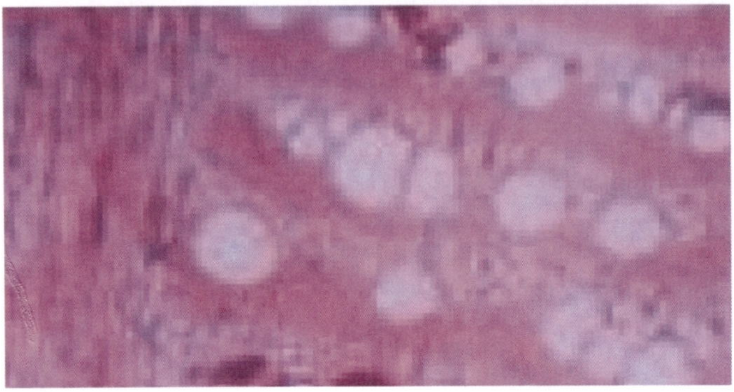

Fig. 13.23: Cross Section of *Grevillea robusta*

24. *Tectona grandis*

Trade Name: Teak

Vernacular Names: Jati, Tek (Indonesia), Java teak (Germany), Kyun (Myanmar), Teca (Brazil), Segan (Beng.), Saga, Sagach (Guj.), Sagon, Sagwan (Hind.), Tega (Kan.), Sag, Saga (Mar.), Singua (Or.), Tekku, Tekkumaram (Tam.), Adaviteeku and Peedateeku (Tel.)

Botanical Name: *Tectona grandis* Linn.f.

Family Name: Verbenaceae

Origin (Distribution): It is a deciduous tree with a rounded crown which under favourable conditions attains large size with a clean cylindrical bole often becoming buttressed and fluted towards the base.

Tree Description: The tree is large to very large, 25-45 m in height and upto 190 cm in diameter, but in the dry hot areas of Madhya Pradesh, Gujarat and Rajasthan the tree is comparatively smaller, often branchy and much fluted in advanced age. Bark light brown or grey with shallow longitudinal furrows and fibrous, about 0.4- 1.8 cm thick, exfoliating in long thick strips.

General features

- *Tectona grandis* (teak) has golden brown heart wood with black streaks. However, wide variation in colour is present between teak available from different regions.

- The heartwood has an oily feel with a distinct smell resembling old leather.
- Teak is a typical ring porous hard wood. The early wood vessels are large and few, compared to the late wood vessels which are small and numerous.
- Wood is moderately hard and moderately heavy.
- Parenchyma is paratracheal (vasicentric and in broad bands); distinct to the eye in the early wood, forming a continuous zone enclosing the vessel along with initial band of parenchyma delimiting growth.
- **Vessels are filed Colour:** Sapwood and heartwood sharply demarcated.
- Heartwood golden brown when fresh or dark brown on exposure occasionally with black streaks with a waxy feel, lustrous sometimes with white glistening deposit, distinct aromatic odour with the smell of leather.
- Sapwood white, pale yellow or grey and well defined.
- **Weight:** Moderately hard and moderately heavy (650 kg/m^3 at 12 % m. c.)
- **Grain:** Straight, sometimes wavy.
- **Texture:** Coarse

Gross features

- Wood is ring porous. Growth rings are distinct, delimited with early wood vessels enclosed in parenchymatous tissues, less than 1-6/ cm.
- Vessels large in early wood, oval in out line, occasionally filled with tyloses and yellowish-white powdery deposits.
- Vessels medium to small in late wood, mostly solitary or in short radial multiples, round to oval in out line, vessel lines of the early wood zone conspicuous on longitudinal surfaces.
- Parenchyma paratracheal- vasicentric and in broad bands, distinct under the hand lens but distinct to the eye in the early wood forming a continuous zone enclosing the vessel along with initial band of parenchyma delimiting growth rings.
- Rays moderately broad, fairly wide spaced and uniformly distributed; visible to the eye, distinct under the lens.

Uses: A versatile wood. Used extensively for ship and boat building, class I general purpose plywood, cabinet making, interior and exterior joinery, flooring and fine furniture, carving, paneling, turnery, sliced for decorative and face veneers. Teak laboratory fittings and laboratory accessories are logical choice due to the acid and alkalis resistant properties of this timber, it is also used for vats, towers etc. in chemical plants. Building construction poles and cross arms, textile mill accessories, musical instruments, mathematical, engineering and drawing instruments and bus bodies. It is used for doors and windows in house construction, paneling and interior fittings. It is also used for air crafts, marine plywood and block boards. It is an approved timber for table tennis tables, frame of carom boards and general requirements of play ground and park equipment, wooden butter-scoop, butter moulds, wooden boxes for microscopic slides, stand for ammunition explosive boxes and caulking mallets. Teak poles are used for scaffolding, fence posts and for overhead and telecommunication lines.

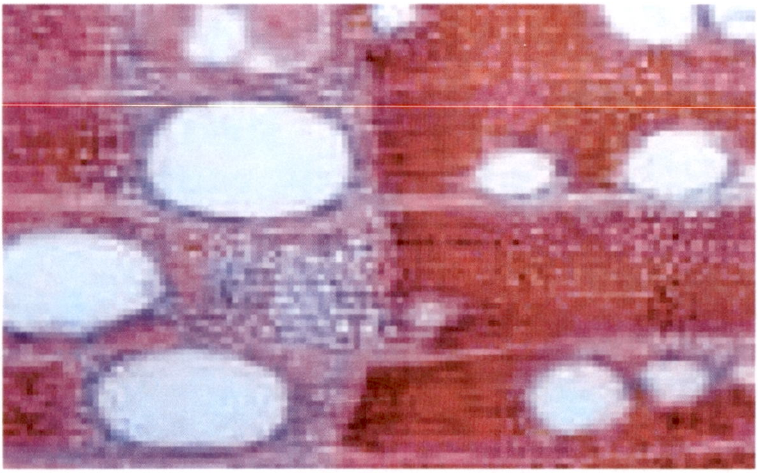

Fig. 13.24: Cross Section of *Tectona grandis*

Appendix

Bibliography

A.C. Dutta: A Class Bookof Botany
E.P Stebbing: Indian Forest insects of commercialimportance.
KA. Chowdry and S.S, Ghosh: Indian Woods Vol. I.
Neison C. Brown: Logging
R.S, Troup: Indian Forest Utilization.
R.S. Ratra: Manual for wood cutters No. 1.
RS. Ratra: Manual for wood cutters No. 2. "Maintenance of saws".
S.S. Negi: A Hand Bookof Forestry
Tribhuwan Mehta: Lecture Notes on Forest Utilization Part I and Part II for Forest Ranger students.